T0076224

CHEMISTRY RESEARCH AND APPLICATIONS

CHEMICAL CRYSTALLOGRAPHY

CHEMISTRY RESEARCH AND APPLICATIONS

Additional books in this series can be found on Nova's website at:

https://www.novapublishers.com/catalog/index.php?cPath=23_29&seriesp=
Chemistry%20Research%20and%20Applications&sort=2a&page=1

Additional e-books in this series can be found on Nova's website at:

https://www.novapublishers.com/catalog/index.php?cPath=23_29&seriespe=
Chemistry+Research+and+Applications

CHEMISTRY RESEARCH AND APPLICATIONS

CHEMICAL CRYSTALLOGRAPHY

BRYAN L. CONNELLY
EDITOR

Nova Science Publishers, Inc.
New York

Copyright © 2010 by Nova Science Publishers, Inc.

All rights reserved. No part of this book may be reproduced, stored in a retrieval system or transmitted in any form or by any means: electronic, electrostatic, magnetic, tape, mechanical photocopying, recording or otherwise without the written permission of the Publisher.

For permission to use material from this book please contact us:
Telephone 631-231-7269; Fax 631-231-8175
Web Site: http://www.novapublishers.com

NOTICE TO THE READER
The Publisher has taken reasonable care in the preparation of this book, but makes no expressed or implied warranty of any kind and assumes no responsibility for any errors or omissions. No liability is assumed for incidental or consequential damages in connection with or arising out of information contained in this book. The Publisher shall not be liable for any special, consequential, or exemplary damages resulting, in whole or in part, from the readers' use of, or reliance upon, this material. Any parts of this book based on government reports are so indicated and copyright is claimed for those parts to the extent applicable to compilations of such works.

Independent verification should be sought for any data, advice or recommendations contained in this book. In addition, no responsibility is assumed by the publisher for any injury and/or damage to persons or property arising from any methods, products, instructions, ideas or otherwise contained in this publication.

This publication is designed to provide accurate and authoritative information with regard to the subject matter covered herein. It is sold with the clear understanding that the Publisher is not engaged in rendering legal or any other professional services. If legal or any other expert assistance is required, the services of a competent person should be sought. FROM A DECLARATION OF PARTICIPANTS JOINTLY ADOPTED BY A COMMITTEE OF THE AMERICAN BAR ASSOCIATION AND A COMMITTEE OF PUBLISHERS.

LIBRARY OF CONGRESS CATALOGING-IN-PUBLICATION DATA

Chemical crystallography / editor, Bryan L. Connelly.
  p. cm.
 ISBN 978-1-60876-281-1 (hardcover)
 1. Crystallography. I. Connelly, Bryan L.
 QD905.2.C44 2009
 548'.3--dc22
                    2010002300

Published by Nova Science Publishers, Inc. ✦ New York

CONTENTS

PREFACE

Chemical crystallography is the study of the principles of chemistry behind crystals and their use in describing structure-property relations in solids. The principles that govern the assembly of crystal and glass structures are described, models of many of the technologically important crystal structures are studied, and the effect of crystal structure on the various fundamental mechanisms responsible for many physical properties are discussed. This new book presents and reviews data on the coordination chemistry of several metal complexes with dipicolinic acid and the crystal structure of some antimalarial metal complexes.

Chapter 1-2,6-Pyridinedicarboxylic acid (dipicolinic acid) is a widely used building block in coordination and supramolecular chemistry. The crystal structure of dipicolinic acid was first solved in 1973, which confirmed its molecular formula of $C_7H_5NO_4$, a molar mass of 167.119 g mol^{-1}, and the resulting composition of its constituent atoms (C, 50.31%; H, 3.02%; N, 8.38%; and O, 38.29%). Dipicolinic acid and its analogues are known to form many intriguing complexes with main group and other metal ions from as far back as 1877. The corresponding bis-acid (DPA), bis-ester (DPE), and bis-amide (DPAM) derivatives behave as tridentate ligands, which efficiently coordinate to various metal ions. This chapter will discuss the coordination chemistry of several metal complexes with dipicolinic acid, its analogues, and derivatives as ligands.

Chapter 2- Transition metal dithiolenes are versatile complexes capable of a wide range of oxidation states, coordination geometries, and magnetic moments.[1] As a consequence, these complexes have been widely studied as building blocks for crystalline molecular materials. Particularly successful are the square-planar metal dithiolenes (Chart 1), from which materials have been produced that exhibit conducting, magnetic, and nonlinear optical properties, as well as superconductivity in some cases.[1-3] In their application to molecular- based conductors, metal dithiolenes can play several different roles. Metal dithiolenes may form an effective conduction pathway through intermolecular face-to-face stacking, or can play a supportive role as counterions to other planar molecules (such as perylene or tetrathiafulvalene derivatives, Chart 1) which provide the actual conduction path.[3] When acting as counterions, such dithiolene complexes can additionally impart magnetic properties to the molecular conductors via interactions between localized spins of the metal dithiolene with the intinerant spins of conduction electrons.

As a result of such research efforts, there now exists a variety of available dithiolene ligands which have been applied to produce a broad range of metal dithiolene materials. One focus of study has been the use of electronically delocalized dithiolene ligands to explore the influence on the solid-state structures and the resulting material properties.[2] Of particular interest has been the preparation and study of metal dithiolenes functionalized with thiophene moieties. The application of such extended π-systems and sulfur-rich ligands are expected to enhance solid-state interactions, which could result in enhanced electrical conductivity or higher magnetic transition temperatures. As the molecular packing in the crystal is determined by the total balance of many weak intermolecular forces (hydrogen bonding, van der Waals, π–π interactions, and S···S/M···S interactions),[2,3] the additional thiophene content would increase such intermolecular interactions and provide more significant overlap of frontier orbitals. In addition, such complexes could provide potential precursors to metal-dithiolene-containing conjugated polymers.

Chapter 3- This paper reviews the crystal chemistry of 25 diphenylguanidine/diphenylguanidinium compounds. Diphenylguanidine is an atropisomer and several conformations have been isolated in the solid state. Such conformations are investigated and the conformation description, chirality, aromaticity of the flexible molecule are systematized in this paper. The dipolar moment and octupolar character are probed. Intermolecular interactions are classified. Two new salts are reported: N,N'-Diphenylguanidinium nicotinate hydrate and N,N'-Diphenylguanidinium 5-nitrouracilate dihydrate. The former crystallizes in a chiral space group, with the phenyl rings of the cation oriented like the blades of a propeller. The latter crystallizes in the centrosymmetric, triclinic space group P-1, and the cation exhibits and *anti-anti* conformation.

Keywords: atropisomer; polarizability; X-ray diffraction.

Chapter 4- Remarkable progress has been achieved in the area of metal-organic frameworks (MOFs) in recent years not only due to their diverse topology and intriguing structures but also owing to their interesting physical and chemical properties. MOFs with specific ion exchange property have attracted great attention for their potential application in molecular/ionic recognition and selective guest inclusion. Despite the difficulty in predicting the structure and property of MOFs, the increasing knowledge regarding the synthesis methods and characterization techniques has largely expanded for the rational designs. In this chapter the recent works in cation/anion exchange with zero- (0D), one- (1D), two- (2D) and three-dimensional (3D) frameworks from our and other groups will be highlighted. Cation exchange mainly concentrates on the metal ions and organic cations such as M^{m+}, $[M(H_2O)_n]^{m+}$, $[Me_2NH_2]^+$, etc., while anion exchange comprises the majority of the counteranions, e.g., ClO_4^-, NO_3^-, BF_4^-, and so on. The functions of the exchanged compounds, i.e., enhancement of gas adsorption and photoluminescence, were greatly reformed.

Chapter 5- The coordination polymer $[Zn(tmbdc)(dmso)_2]\cdot2(DMSO)$ (tmbdc = 2,3,5,6-tetramethyl-1,4-benzenedicarboxylate) has been synthesized by layer diffusion in DMSO (dimethyl sulfoxide) solution. The compound contains 1D chain formed by octahedraly coordinated Zn^{2+} ion chelated by the carboxyl groups of tmbdc. In another recently reported coordination polymer $[Zn_2(bdc)_2(dmso)_2]\cdot5(DMSO)$ (bdc = 1,4-benzenedicarboxylate) prepared under the same condition, pairs of Zn^{2+} ions are bridged by four carboxyl groups to form paddle-wheel sub unit and the 2D (4,4) net structure. Analysis of the structures reveals that the substituents of the ligands determine the coordination environments of zinc ions and the coordination modes of the carboxyls, and thus the final structures of the coordination polymers.

In: Chemical Crystallography   ISBN: 978-1-60876-281-1
Editors: Bryan L. Connelly, pp. 1-68   © 2010 Nova Science Publishers, Inc.

Chapter 1

DIPICOLINIC ACID, ITS ANALOGUES, AND DERIVATIVES: ASPECTS OF THEIR COORDINATION CHEMISTRY

Alvin A. Holder,[1] Lesley C. Lewis-Alleyne,[1]*
Don vanDerveer,[2] and Marvadeen Singh-Wilmot[3]
[1]The University of Southern Mississippi, Department of Chemistry and
Biochemistry, 118 College Drive, Box # 5043, Hattiesburg, MS 39406.
[2] Molecular Structure Center, Chemistry Department,
Clemson University, Clemson, SC 29634-0973.
[3] Department of Chemistry, The University of the
West Indies, Mona Campus, Mona, Kingston 7, Jamaica.

ABSTRACT

2,6-Pyridinedicarboxylic acid (dipicolinic acid) is a widely used
building block in coordination and supramolecular chemistry. The crystal
structure of dipicolinic acid was first solved in 1973, which confirmed its
molecular formula of $C_7H_5NO_4$, a molar mass of 167.119 g mol^{-1}, and the
resulting composition of its constituent atoms (C, 50.31%; H, 3.02%; N,

* Corresponding author: E-mail: alvin.holder@usm.edu Telephone: 601-266-4767, Fax: 601-
266-6075.

8.38%; and O, 38.29%). Dipicolinic acid and its analogues are known to form many intriguing complexes with main group and other metal ions from as far back as 1877. The corresponding bis-acid (DPA), bis-ester (DPE), and bis-amide (DPAM) derivatives behave as tridentate ligands, which efficiently coordinate to various metal ions. This chapter will discuss the coordination chemistry of several metal complexes with dipicolinic acid, its analogues, and derivatives as ligands.

1.0. INTRODUCTION

2,6-Pyridinedicarboxylic acid (dipicolinic acid), **I**, is a widely used building block in coordination and supramolecular chemistry.[1-5] It is a versatile, strong, nitrogen-oxygen, multi-modal donor ligand, which forms stable complexes with diverse metal ions, sometimes in unusual oxidation states, for example, its corresponding bis-acid (DPA), bis-ester (DPE), and bis-amide (DPAM) derivatives behave as tridentate ligands, which efficiently coordinate to various metal ions.[6-13]

I

The crystal structure of dipicolinic acid was first solved by Takusagawa *et al.*[14] in 1973. Takusagawa *et al.*[14] confirmed its molecular formula of $C_7H_5NO_4$, a molar mass of 167.119 g mol^{-1}, and the resulting composition of its constituent atoms (C, 50.31%; H, 3.02%; N, 8.38%; and O, 38.29%).

More recently, (creatH)$^+$(Hdipic)$^-$.H$_2$O was synthesized by the reaction between dipicolinic acid and creatinine (creat) (Scheme 1).[15] Its structure consists of (creatH)$^+$ and (Hdipic)$^-$ ions and a disordered water molecule (Figure 1), all lying on a crystallographic mirror plane.[15] It was reported that the intermolecular interactions among these three fragments consist of ion-pairing, hydrogen bonding and π-π stacking. A single proton transfer occurs from one of the two carboxylic acid functional groups to the endocyclic imine N atom of creatinine. This results in the localization of the exocyclic C8-N4 double bond [1.300 (2) Å] and the adjacent single bond C8-N3 [1.369 (2) Å].

These values can be compared with the intermediate, delocalized values in the parent neutral creatinine molecule [1.320 (3) and 1.349 (3) Å, respectively].[16] The two carboxylic groups of the (Hdipic)⁻ anion adopt slightly different conformations, both being essentially coplanar with the pyridine ring. It was reported that all of the N and O heteroatoms participate in extensive strong or weak hydrogen-bonding interactions, particularly the strong O3•••O2^i interaction.

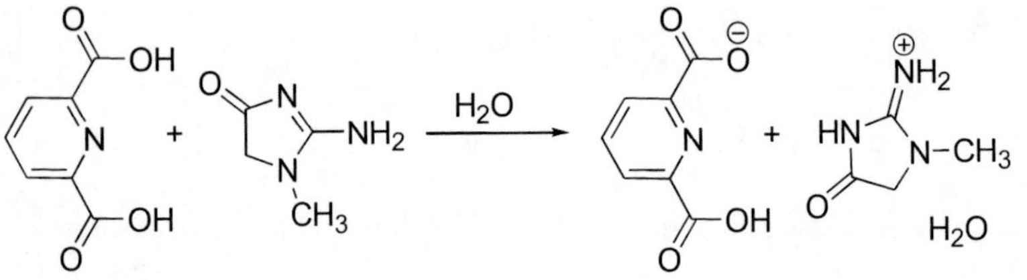

Scheme 1

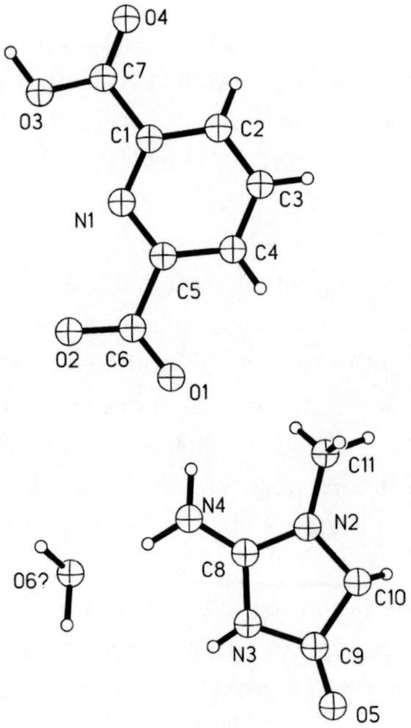

Figure 1. A diagram of (creatH)$^+$(Hdipic)⁻.H$_2$O. (Reproduced by permission from reference 15)

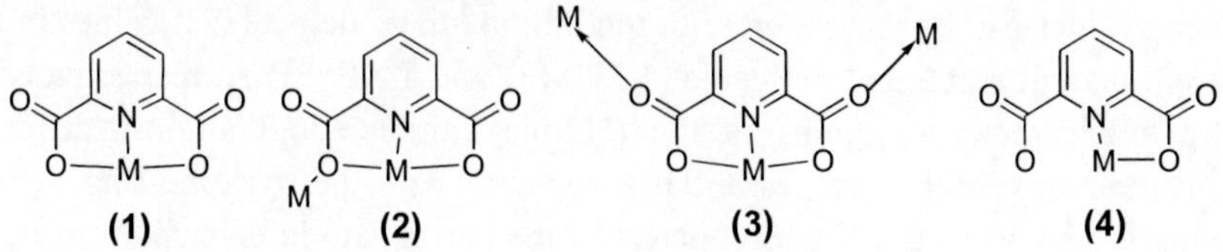

Figure 2. Different coordination modes of the dipicolinate anion.

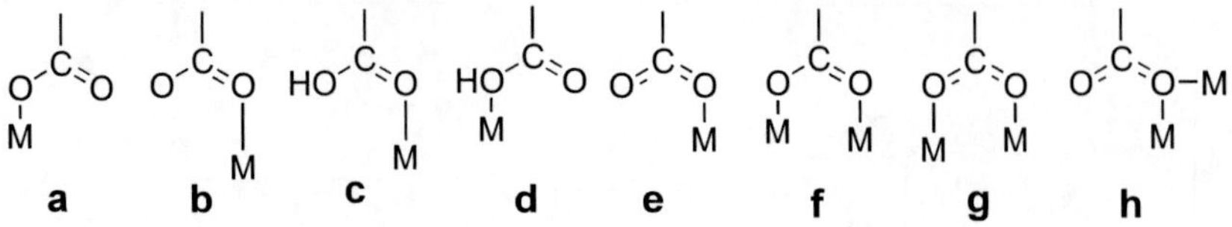

Figure 3. Coordination modes observed in the solid state of metal-carboxylate complexes.

Dipicolinic acid and its analogues are known to form many intriguing complexes with main group and transition metal ions from as far back as 1877.[17] Since the discovery of the dipic^{2-} ligand in a biological system,[18] its coordination chemistry has been extensively investigated. Several modes of coordination are known: O-unidentate[19] H$_2$dipic; O,N-bidentate[20] Hdipic⁻; O,N,O-tridentate[19, 21-26] H$_2$dipicH, Hdipic⁻, and dipic^{2-}; (O,N)O-bidentate bridging[27, 28] dipic^{2-}; (O,N,O)O and O,N,i-O-tridentate bridging[23, 25, 29-33] dipic^{2-}. There has been renewed interest in the complexes of this ligand from several standpoints including unconventional physical properties such as liquid crystal behavior and nonlinear optics,[26, 28] DNA cleavage,[34] electron transfer,[35-39] activation of dioxygen,[40-48] and novel coordination modes.[28]

Different coordination modes have been reported in transition metal-dipicolinate complexes.[28] These are shown in Figure 2:

The coordination modes observed in the solid state of metal-carboxylate complexes are shown in Figure 3. The coordination modes previously reported are labeled a,[23, 25, 49-54] c,[22, 23, 25, 49-55] e,[22, 52, 54] f,[23] g,[51] and h.[25] Coordination modes b and d have been reported in the literature.[56]

The most common coordination mode is **a** in which the metal is coordinated to the long C-O bond.[56] This coordination mode is found in [Ni(Hdipic)$_2$].3H$_2$O,[49, 50, 53] [Zn(Hdipic)$_2$].3H$_2$O,[51, 52] [Zn$_2$(dipic)$_2$].7H$_2$O,[51] [Fe(Hdipic)$_2$(OH$_2$)],[25] [Fe$_2$(dipic)$_2$(OH$_2$)$_6$]. 2H$_2$dipic,[25] [Fe$_3$ (dipic)$_2$ (Hdipic)$_2$ (H$_2$O)$_4$].2H$_2$O,[23] [Fe$_2$ (dipic)$_2$ (H$_2$O)$_5$] .2.25H$_2$O ,[23] [Fe$_3$ (dipic)$_4$ (H$_2$O)$_6$ (NH$_4$)$_2$].4 H$_2$O. 2H$_2$dipic,[23] [Fe$_{13}$ (Hdipic) $_6$(dipic)$_{10}$ (H$_2$O)$_{24}$]. 13H$_2$O ,[23]

[Cu(dipic) (H$_2$O)$_2$],[54] and [Cu(H$_2$d ipic)(d ipic)].H$_2$O.[52] Another common coordination mode found in complexes with monoprotonated Hdipic⁻ is represented by **c** and found in [Ni (Hdipic)$_2$].3H$_2$O,[49, 50, 53] [Zn(Hdipic)$_2$]. 3H$_2$O,[51, 52] [Fe (Hdipic)$_2$ (OH$_2$)], 20 [Fe$_3$ (dipic)$_2$ (Hdipic)$_2$(H$_2$O)$_4$].2H$_2$O,[23] [Fe$_{13}$(Hdipic)$_6$(dipic)$_{10}$(H$_2$O)$_{24}$].13H$_2$O,[23] and [Cu(H$_2$dipic)(dipic)].H$_2$O.[52, 55] A coordination mode in which the C-O bonds are of equal length is **e**, which has been observed in [Ag(Hdipic)$_2$].H$_2$O,[22] [Cu(H$_2$dipic)(dipic)].3H$_2$O,[52, 55] and [Zn(Hdipic)$_2$].3H$_2$O.[52] Less commonly observed coordination modes involve coordination of two metal ions to one carboxylate group and are represented by **f** (observed in [Fe$_3$(dipic)$_2$ (Hdipic)$_2$ (H$_2$O)$_4$] .2H$_2$O, [23] [Fe$_2$ (dipic)$_2$(H$_2$O)$_5$].2.25H$_2$O,[23] and [Fe$_{13}$(Hdipic)$_6$(dipic)$_{10}$(H$_2$O)$_{24}$].13H$_2$O),[23] **g** (observed in [Zn$_2$ (dipic)$_2$]. 7H$_2$O),[51] and **h** (observed in [Fe$_2$ (dipic)$_2$ (OH$_2$)$_6$]. 2H$_2$dipic).[25] The key differences in the two new coordination modes are the coordination of the metal ion to the short C=O bond (in **b**) and to the oxygen atom carrying the proton (in **d**).

More recently, there was a report on the use of dipicolinic acid in the design of layered crystalline materials using coordination chemistry and hydrogen bonds. MacDonald *et al.*[57] reported the synthesis and characterization of several first-row transition metals with dipicolinic acid as a ligand. Five bis(imidazolium 2,6-pyridinedicarboxylate)M(II) trihydrate complexes (where M = Mn^{2+}, Co^{2+}, Ni^{2+}, Cu^{2+}, or Zn^{2+}), were synthesized from the reaction between dipicolinic acid and imidazole with Mn^{2+}, Co^{2+}, Ni^{2+}, Cu^{2+}, or Zn^{2+} salts.[57]

2.0. COORDINATION CHEMISTRY OF DIPICOLINIC ACID AND ITS ANALOGUES

This chapter discusses the coordination chemistry of selected main group and transition metal complexes with dipicolinic acid, its analogues, and derivatives as ligands. Selected elements will be presented in terms of increasing atomic number. Out of all of the alkali metals, there has been a report of the crystal structure of sodium coordinated to dipicolinic acid.[58] Calcium, magnesium, and strontium, three alkaline earth metals, are popular metal centers, which have been reported in the literature to be coordinated to dipicolinic acid or its analogues.[32, 33, 59-62]

The structure of the deep red diaquoperoxotitanium(IV) dipicolinate complex, [TiO$_2$(C$_7$H$_3$O$_4$N)(H$_2$O)$_2$].2H$_2$O was reported.[63] The complex (see

6 Alvin A. Holder, Lesley C. Lewis-Alleyne, Don vanDerveer et al.

Figure 4) has a pentagonal bipyramidal seven-fold coordination with two carboxylate oxygens, one nitrogen and two oxygens of the peroxo group forming a distorted pentagon and two water oxygens at the apices. The peroxo group is attached laterally to the titanium(IV) metal center. It was reported that the pentagon is virtually planar, with the distances of Ti from the least square plane being less than standard deviation.[63]

It was reported that the O-O distance in the peroxide is 1.458 Å, and that this value agreed well with values of 1.464, 1.463, and 1.469 Å in the triclinic diaquo, difluoro, and nitrilotriacetic acid (NTA) complexes, respectively.[63] The Ti-O_{peroxo} distance (1.833 Å) was compared with the values 1.834 and 1.856 Å, 1.846 and 1.861 Å, and 1.889 and 1.892 Å, respectively, and the Ti-O_{water} apical distances of 2.018 Å with the Ti-O (F) values of 2.022 and 2.055 Å in the triclinic diaquo, 1.853 and 1.887 Å in the difluoro and 1.819 and 2.065 Å in the NTA complexes.[63] It was concluded that while the O-O bond distance of the peroxo group was practically the same in all the structures, there was a small but significant variation in the Ti-O_{peroxo} and apical distances.[63] There was progressive increase in the Ti-O_{peroxo} bond lengths down the respective series, and a corresponding decrease in the apical bond lengths.[63]

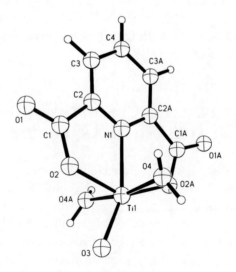

Figure 4. A diagram of [TiO_2(C_7H_3O_4N)(H_2O)_2].2H_2O. (Reproduced by permission from reference 63)

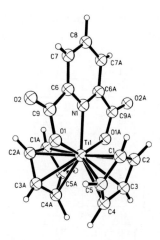

Figure 5. A diagram of [(C₅H₅)₂Ti(dipic)]. (Reproduced by permission from reference 64)

Reaction of $(C_5H_5)_2Ti(CH_3)_2$ or $(CH_3)_4C_2(C_5H_4)_2Ti(CH_3)_2$ with dipicolinic acid produced several titanocene dipicolinate derivatives.[64] Figure 5 shows the structure of one of those derivatives. As expected from structural studies on other transition metal dipicolinate complexes, [19, 21-26] the dipicolinate ligand is bound to the Ti(IV) metal centre by its pyridine N atom and two of the carboxylate O atoms, which occupy the central and the two lateral coordination sites of the titanocene fragment. The Ti-N and Ti-O distances of 216 and 211 ppm were reported to be significantly longer than Ti-N≡ bonds (196-202 pm)[65-67] and Ti-O bonds (186-190 pm)[68-70] in comparable, tetracoordinate titanocene complexes. The O-Ti-N angle of 71.1° was reported to be within the range of 65-73° found in other pentacoordinate, non-hydridic metallocene derivatives.[71, 72]

The titanium(IV) metal center and its N an O ligand atoms are coplanar by crystal symmetry; the TiO(1)NO(1') plane is perpendicular to the ring centroid-Ti-centroid plane; the two planes intersect at an angle of 89.9°; the plane of the pyridine ring is not quite coplanar with the TiO(1)NO(1) plane. A slight rotational deviation of these two planes by 3.6° is connected with a rotation of both CO_2 groups by 4.8° out of plane of the pyridine ring, and by 7.4° out of the TiO(1)NO(1') plane. A similar deviation from coplanarity, which was reported to have led to a slight shortening of the Ti-N relative to the two Ti-O distances, with respect to a fully coplanar geometry, has been reported for the Ti(IV) complex, $[(H_2O)_2(O_2)Ti(dipic)]$.[63, 73]

Table 1. The bond lengths (pm) and bond angles (°) at the Ti(IV) metal centre in [(C₅H₅)₂Ti(dipic)].

Ti-O	211.1(6)	O-Ti-N	71.1(2)
Ti-N	216.0(8)	CR-Ti-CR'	133.0
Ti-CR	205.2	PL-PL'	47.2
Ti-PL	205.2		
Ti-C(1)	238(1)		
Ti-C(2)	237(1)		
Ti-C(3)	235(1)		
Ti-C(4)	238(1)		
Ti-C(5)	236(1)		

CR = centroid of C_5 ring, PL = mean plane of C_5 ring.

For the complex, $[(C_5H_5)_2Ti(dipic)]$ (Figure 5), Leik et al.[64] concluded that it is apparent that the dipicolinate ligand, with its rather small bite angle of ~70°, is almost ideally suited to induce a pentacoordinate geometry even at a $(C_5H_5)_2Ti$ centre, which otherwise appears to avoid this increase in coordination number, probably for steric reasons. Table 1 shows the bond lengths and bond angles at the Ti(IV) metal centre in $[(C_5H_5)_2Ti(dipic)]$.

Vanadium, in different oxidation states, has been used in conjunction with dipicolinic acid and its analogues to produce coordination complexes.[62, 74-90] A selection of vanadium-containing complexes is discussed below.

The novel complex, $C(NH_2)_3[VO_2(dipic)].2H_2O$, and its analogous complex, $NH_4[VO_2(dipic)]$, were synthesized and characterized.[77] Figure 6 shows ORTEP diagrams for both anions. The vanadium(V) metal center shows a similar pentacoordinated environment in the guanidinium and ammonium salts of $[VO_2(dipic)]^-$.[77] In the anion, the VO_2^+ group is coordinated to a $dipic^{2-}$ acting as a tridentate ligand through its carboxylic oxygen atoms [V–O distances of 1.983(2) and 1.988(2) Å ° in $C(NH_2)_3[VO_2(dipic)].2H_2O$ and 1.974(2) and 1.978(2) Å in $NH_4[VO_2(dipic)]$] and the nitrogen atom [V–N distances of 2.086(2) (for $C(NH_2)_3[VO_2(dipic)].2H_2O$) and 2.091(2) Å (for $NH_4[VO_2(dipic)]$)]. The $dipic^{2-}$ ligand is planar [rms deviation of atoms from the least-squares plane of 0.012 (for $C(NH_2)_3[VO_2(dipic)].2H_2O$) and 0.051 Å (for $NH_4[VO_2(dipic)]$)] with the metal ion lying onto this plane in $C(NH_2)_3[VO_2(dipic)].2H_2O$ [at 0.001(7) Å] and slightly above in $NH_4[VO_2(dipic)]$ [at 0.136(1) Å]. The ligand plane of $C(NH_2)_3[VO_2(dipic)].2H_2O$ bisects the dioxovanadium V=O double bonds whose lengths are 1.614(7) and 1.626(7) Å. This agrees with the structural data reported for the [VO_2(dipic)]$^-$ complex in Cs[VO_2(dipic)].H_2O where the V=O double bond distances are 1.610(6) and 1.615(6) Å.[91] In contrast, the ligand plane of NH_4[VO_2(dipic)] structure departs appreciable from the O1=V=O2 bisector and the V–O1 bond distance is 0.012 Å (i.e. six times the rms error) longer than the V–O2 length [1.612(2) Å]. That difference in the VO distances is probably due to a pair of medium to strong N–H•••O1 bonds with the NH_4^+ counter-ion (see below). There was a report of an example of even more pronounced V=O bond asymmetry in the VO_2^+ group, namely in a bis oxo bridged binuclear vanadium(V) complex of stoichiometry [$CH_3NHC(NH_2)_2$]$_2$[V_2O_4(dipic)$_2$].[80] In this compound, the oxygen atom of the VO_2^+ moiety, laying near the coordination plane, bridges the two halves of a centro-symmetric dimer through a weak axial V•••O bond. As a consequence of this, the two V=O distances differ in 0.078(1) Å [d(V–O1) = 1.606(1) Å].[77]

In C(NH_2)$_3$[VO_2(dipic)].$2H_2O$, the planar guanidinium counterion lays parallel to the dipic^{2-} ligand at a van der Waals contact distance of 3.1 Å. [77] The [VO_2(dipic)]$^-$ and [C(NH_2)$_3$]$^+$ ions are arranged in the lattice along layers parallel to (010) crystallographic planes. These layers are stabilized by a network of medium to strong intra-layer N–H•••O bonds involving the guanidinium NH_2 groups and carboxylic oxygen atoms of the dipic^{2-} ligand [N..O distances and N–H•••O angles are found in the ranges 2.848–2.982 Å and 132.6–176.1°, respectively]. Two parallel N–H•••O bonds are formed between the N2 and N3 atoms of the guanidinium ion and the carboxylate oxygens O3 and O4 of a neighboring anion. The interaction gives rise to an hexagonal pattern analogous to the observed in the bis-oxo-bridged V(V) complex[80] and similar to that found in a large number of H-bonded layered crystals and to that described for the interaction of phosphate, sulfonate, carboxylate and nitrate with this cation.[77] The N1 nitrogen atom of guanidinium interacts, through one of its hydrogen atoms, with the O6 oxygen from a carboxylate ion of another neighboring unit. The lattice is further stabilized by inter-layer H-bonds mediated by one crystallization water molecule held onto the layer by a N–H•••Ow bond [the N•••Ow distance and the N–H•••Ow angle are 2.825 Å and 157.5°, respectively]. This molecule is acting as a donor in two Ow–H•••O interactions with the dioxovanadium groups of neighboring layers [the corresponding Ow•••O distances and Ow–H•••O angles are 2.823 and 2.860 Å and 152.6° and 145.4°, respectively].

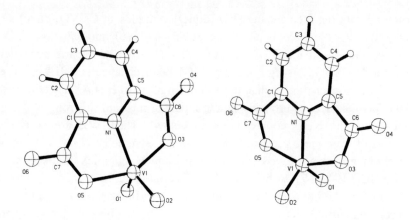

Figure 6. Diagrams of two [VO$_2$(dipic)]⁻ anions. (Reproduced by permission from reference 77)

[V^{IV}O(H$_2$O)$_2$(dipic)].2H$_2$O was synthesized by the reaction of VO(acac)$_2$ with dipicolinic acid.[85] The X-ray three-dimensional structural determination of [V^{IV}O(H$_2$O)$_2$(dipic)].2H$_2$O revealed that it crystallizes in the triclinic space group P_11 with two molecules in the unit cell, and consists of [V^{IV}O(H$_2$O)$_2$(dipic)] and two lattice water molecules. As illustrated in Figure 7, the geometry is a disordered octahedron with vanadium(IV) coordinated by two oxygen atoms from water, two carboxyl oxygen atoms (COO) and one nitrogen atom from a dipicolinate ligand and one terminal oxo[85]. The dipicolinate ligand (COO, COO, N) chelates one vanadium atom to form two five-membered rings. This was referred as a new example of a vanadium(IV) complex with dipicolinate, different from other related vanadium complexes, e.g., potassium oxodiperoxo(pyridine-2-carboxylate)vanadate(V), K$_2$[VO(O$_2$)$_2$(PA)].2H$_2$O;[92] potassium oxodiperoxo (3-hydroxypyridine-2-carboxylate)vanadate (V), K$_2$[VO(O$_2$)$_2$ (3HPA)]. 3H$_2$O;[92] K$_3$ [VO(O$_2$)$_2$ (2,4-pyridinedicarboxylate)].2H$_2$O;[93] bpV (2,4-pdc); K$_3$ [VO(O$_2$)$_2$ (3-acetatoxypicolinate)] .2H$_2$ O, bpV(3-acetpic);[93, 94] [VO(3HPA) (H$_2$O)]$_4$.9H$_2$O,[94] V(pic)$_3$. H$_2$O,[95] and [VO(6epa)$_2$ (H$_2$O)]. 4H$_2$O. [85] For [V^{IV}O (H$_2$O) $_2$(dipic)]. 2H$_2$O, the V–O(1) bond length [1.594 (3) Å] is shorter than those in K$_2$[VO(O$_2$)$_2$ (PA)]. 2H$_2$O, K$_2$[VO(O$_2$)$_2$ (3HPA)].3H$_2$O, bpV(2,4-pdc), bpV(3-acetpic), [VO(3HPA) (H$_2$O)]$_4$.9H$_2$O, V(pic)$_3$ .H$_2$O and [VO(6epa)$_2$ (H$_2$O)].4H$_2$O, while the V–N bond length [2.163(4) Å] is slightly longer than those in the corresponding complexes above. The V–O$_{carb}$ distances in [V^{IV}O(H$_2$O)$_2$(dipic)].2H$_2$O are shorter those of K$_2$[VO(O$_2$)$_2$(PA)].2H$_2$O, K$_2$[VO(O$_2$)$_2$(3HPA)].3H$_2$O, bpV(2,4-pdc) and pbV(3-acetpic), close to that of

[VO(3HPA)(H$_2$O)]$_4$.9H$_2$O and longer than those in V(pic)$_3$.H$_2$O and [VO(6epa)$_2$(H$_2$O)].4H$_2$O. However, it is surprising that the V–O$_{water}$ distances [2.016(4), 2.059(4) Å] are much shorter than those in the corresponding complexes.

The O=V–N and O=V–O$_{carb}$ angles are 178.04(11)° and 106.43(16)° in [V^{IV}O(H$_2$O)$_2$(dipic)].2H$_2$O, respectively.[85] Both angles are different from those in related vanadium complexes, perhaps because coordinated nitrogen is trans to a terminal oxygen atom in [V^{IV}O(H$_2$O)$_2$(dipic)].2H$_2$O, while, for other corresponding vanadium complexes, the coordinated O (COO) atom is trans to the terminal oxygen atom. The N–V–O$_{carb}$ angle is close to those found in other vanadium complexes. Comparisons of the detailed bond distances and angles related to vanadium complexes are given in Tables 2 and 3, respectively.[85]

It is worth noting that C(8)–O(7) and C(8)–O(4), and C(7)–O(5) and C(7)–O(6) are shortened, indicating partial double bond character. C(8)–O(7) and C(8)–O(4) are 1.232(3) and 1.289(4) Å,[85] respectively; while C(7)–O(5) and C(7)–O(6) are 1.298(4) and 1.227(4) Å, respectively, indicating more electron delocalization in [V^{IV}O(H$_2$O)$_2$(dipic)].2H$_2$O than in V(pic)$_3$.H$_2$O.[95]

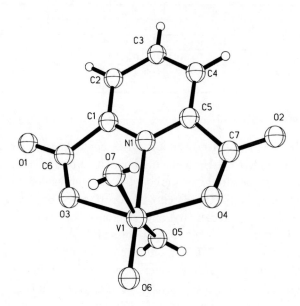

Figure 7. A diagram of [V^{IV}O(H$_2$O)$_2$(dipic)]. (Reproduced by permission from reference 85)

Table 2. Comparison of the bond lengths (Å) in the related complexes.

Complex	V=O	V-N	V-O$_{carb}$	V-O$_{water}$	Reference
[V^{IV}O(H$_2$O)$_2$(dipic)].2H$_2$O	1.594(3)	2.163(4)	2.026(3)-2.051(4)	2.016(4)-2.059(4)	[85]
K$_2$[VO(O$_2$)$_2$(PA)	1.599(4)	2.123(5)	2.290(4)		[96]
K$_2$[VO(O$_2$)$_2$(3HPA)].3H$_2$O	1.606(2)	2.137(2)	2.314(2)		[96]
bpV(2,4-pdc)	1.622(9)	2.144(11)	2.299(8)		[92]
bpV(3-acetpic)	1.621(3)	2.179(4)	2.190(6)		[92]
[VO(3HPA)(H$_2$O)]$_4$.9H$_2$O	1.584-1.608	2.124-2.152	1.963-2.154	2.034-2.073	[93]
V(pic)$_3$.H$_2$O		2.112(3)-2.153(3)	1.936(3)-1.966(2)		[94]
[VO(6epa)$_2$(H$_2$O)].4H$_2$O	1.572(6)-1.596(6)	2.118(5)-2.153(5)	1.956(5)-2.002	2.219(5)-2.283(5)	[95]

Table 3. Comparison of the angle (°) in the related complexes.

Complex	N-V=O	O=V-O$_{carb1}$	O=V-O$_{carb2}$	N-V-O$_{carb}$	Reference
[V^{IV}O(H$_2$O)$_2$(dipic)].2H$_2$O	178.07(11)	106.40(17)	82.27(16)	73.39(15)-73.75(15)	[85]
K$_2$[VO(O$_2$)$_2$(PA)	93.6(2)	166.7(2)		73.1(2)	[96]
K$_2$[VO(O$_2$)$_2$(3HPA)].3H$_2$O	94.92(7)	168.73(7)		73.0(6)	[96]
bpV(2,4-pdc)	93.1(4)	166.3(4)		73.7(6)	[92]
bpV(3-acetpic)	93.39(16)	166.04(15)		72.7(4)	[92]
[VO(3HPA)(H$_2$O)]$_4$.9H$_2$O	91.9-95.9	158.2-160.7	96.7-98.4	73.3-90.1	[93]
V(pic)$_3$.H$_2$O				76.48(10)-168.58(10)	[94]

In addition, [V^{IV}O(H$_2$O)$_2$(dipic)].2H$_2$O contains some hydrogen bonds, primarily Oligand(H$_2$O)–H•••Ouncoordinated carbonyl, Oligand(H$_2$O)•••H–Ow and Ow–H•••Ocoordinated carbonyl.[85] The molecules are linked by two types of hydrogen bond. One is between coordinated oxygen atoms from the water and uncoordinated carbonyl oxygen atoms from dipicolinate ligands, the others are bridging hydrogen bonds formed between coordinated oxygen atoms from the water and coordinated carboxyl oxygen atoms from dipicolinate ligands by lattice water (Ow1) being bridged, namely Ooxo [O(2)]–H•••Ow(1)–H•••Ocoordinated carbonyl [O(4)]. Mononuclear vanadium [V^{IV}O(H$_2$O)$_2$(dipicolinate)].2H$_2$O units and crystallization water molecules are held together in an extensive two-dimensional network via O–H•••O hydrogen bonds, π–π stacking interactions between parallel aromatic pyridines, and face-to-face stacking interactions between parallel carboxylate groups of dipicolinate along the plane formed by the x, z axis of the unit cell. The V–V distance between molecules along the z axis is 6.568 Å ; along the x axis it is 9.129 Å.

[VO(dipic)(phen)].3H$_2$O was synthesized and characterized by X-ray crystallography.[83] The deprotonated dipicolinic acid acting as a tridentate chelating agent coordinates to the V(IV) metal centre through the heterocyclic ring nitrogen N(1) and the carboxylate oxygens O(3) and O(4). All three of them occupy three positions of a distorted square plane (Figure 8), the fourth position being occupied by one of the phen nitrogen atoms N(3).[83] The oxygen atom of the vanadyl moiety lie above the plane defined by O(3)-N(1)-O(4)-N(3), while the position *trans* to the vanadyl oxygen is occupied by the N(2) nitrogen atom of the coordinated phen ligand. The V-O(5) distance of 1.581 (3) Å is a little shorter than is generally found in most V(IV) complexes with a nitrogen donor attached to its trans position.[97, 98] Of the two V—N bonds generated by the coordinated phen ligand, the V--N(2) bond *trans* to the V=O bond is longer (2.312 Å) than the other V-N(3) bond (2.126 Å). The vanadium(IV) metal centre exists in a distorted octahedral donor environment. The deviation of the vanadium atom from the plane defined by O(3)-N(1)-O(4)--N(3) is 0.2910 Å and the dihedral angle between the mean planes defined by the aromatic ligands is 93.90°. Figure 1 shows that O(lw), O(2w) and O(3w) form part of the asymmetric unit. O(lw) is directly H-bonded with O(1) of the molecule. Each of the O(lw), O(2w) and O(3w) are connected to the symmetry generated O(lb), O(2a) and O(4c), respectively with transformation codes a(x,1 + y, z); b(1 - x, 1 - y, -z); c(x, 0.5 - y, - 0.5 - z) . Two of the oxygens O(lw) and O(2w) form three hydrogen bonds whereas O(3w) is connected by only two hydrogen bonds. Molecules are packed within the lattice through this type of hydrogen bonds. O(lw) exhibits two-fold disorder and accordingly atom O(lw) and O(1 'w) are assigned 0.7 and 0.3 occupancy, respectively.[83]

4-Hydroxypyridine-2,6-dicarboxylatodioxovanadate(V) ihydrate was synthesized and characterized by X-ray crystallography.[99] (NMe$_4$)[VO$_2$(dipic-OH)].H$_2$O contains discrete [VO$_2$(dipic-OH)]⁻ complex anions. The structure of the anion is shown in Figure 9. The asymmetric unit contains two formula units of (NMe$_4$)[VO$_2$(dipic-OH)].H$_2$O, which exhibit only minor structural differences.[99] The vanadium(V) metal center is five coordinate by virtue of coordination by two oxo ligands and the tridentate [dipic-OH]$^{2-}$ ligand

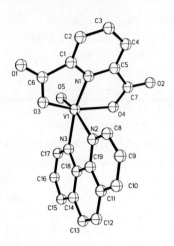

Figure 8. A diagram of [VO(dipic)(phen)]. (Reproduced by permission from reference 83).

(utilizing two carboxylate oxygen atoms and the pyridine nitrogen atom). The hydroxyl group (O(5), O(15)), one carboxylate oxygen atom (O(3), O(13)), and one oxo ligand (O(6), O(16)) from each of the [VO$_2$(dipic-OH)]$^-$ ions in the asymmetric unit form hydrogen bonds to water molecules, resulting in extended chains of [VO$_2$(dipic-OH)]$^-$ anions. The chains are separated by the tetramethylammonium cations.[99] The oxo ligands (O(6), O(16)) that are involved in hydrogen bonding with water form slightly longer bonds to vanadium (1.626(3), 1.627(3) Å) than do the oxo ligands (O(7), O(17)) that do not participate in hydrogen bonding (1.615(3), 1.612(3) Å). The shorter V=O bond lengths are similar to the V=O bond lengths observed in [VO$_2$(dipic)]$^-$ (1.610(6), 1.615(6) Å).[100] Hydrogen bonding to water does not seem to influence significantly the C-O(carboxylate) bond lengths. Other bond lengths and angles in the primary coordination sphere (V-O(carboxylate), V-N(pyridine), and V-O(oxo)) are similar to those observed for [VO$_2$(dipic)]$^-$.[100] An asymmetry in the V-O(carboxylate) bonding is observed; V-O(2) (1.998(4) Å) is slightly shorter than V-O(1) (2.022(3) Å), and a corresponding asymmetry is also observed for the other complex ion in the asymmetric unit.[99]

For Na[VO$_2$(dipic-OH)].2H$_2$O, the structure of the anion is shown in Figure 10. The asymmetric unit contains two formula units of Na[VO$_2$(dipic-OH)].2H$_2$O. As in (NMe$_4$)[VO$_2$(dipic-OH)].H$_2$O, the vanadium(V) atom is five coordinate by virtue of coordination to two oxo ligands and the tridentate [dipic-OH]$^{2-}$ ligand. The Na$^+$ cation is incorporated into a polymeric chain formed by coordination of Na$^+$ by [VO$_2$(dipic-OH)]$^-$ anions. The sodium ion is six-coordinate by virtue of coordination to two water molecules (O(8), O(9)), to two bridging oxo ligands (from two symmetry-related complexes (O(7), O(7B)), and to two carboxylate oxygen atoms from a third symmetry-related complex (O(2A), O(4A) in Figure 10). The carboxylate group at C(7) is therefore in a μ^3 coordination mode, and the group at C(6) is in a terminal monodentate coordination mode. The hydroxyl group O(5) (H-donor), the carboxylate group at C(6) (H-acceptor), and the oxo ligand O(6) (H-acceptor) form hydrogen bonds to the water molecules coordinated to the sodium ion, resulting in a linking together of the polymeric chains into extended sheets. Hydrogen bonding to water and coordination to the sodium ion influences bond lengths within the carboxylate groups.[99] For example, the difference in the C-O bond lengths in the carboxylate at C(6)

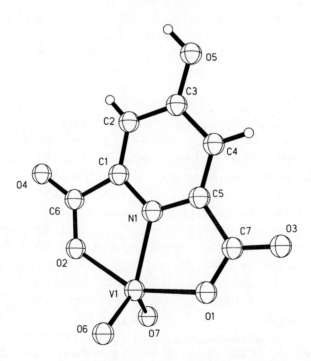

Figure 9. A diagram of the [VO₂(dipic-OH)]⁻ anion. (Reproduced by permission from reference 99).

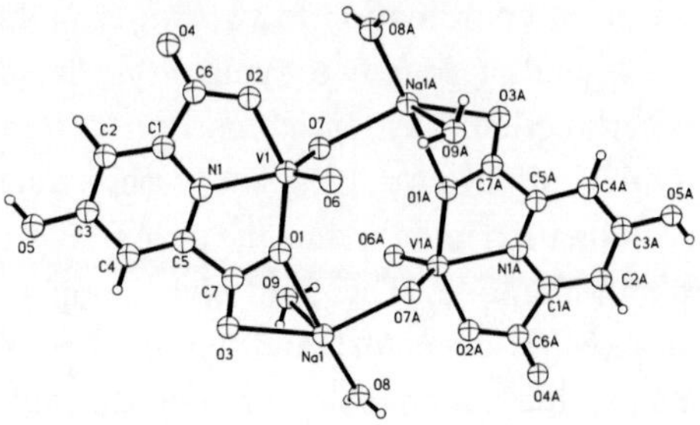

Figure 10. A diagram of the [VO$_2$(dipic-OH)]⁻ anion. (Reproduced by permission from reference 99).

(1.295(3), 1.233(3) Å), which engages in hydrogen bonding to water through O(3), is less pronounced than the difference in the C-O bond lengths seen in the carboxylate at C(7) (1.305(3), 1.223(3) Å), where both of the oxygen atoms coordinate the sodium ion. In contrast, for the oxo ligands hydrogen bonding to water (O(6)) or coordination to sodium (O(7)) does not result in an observable difference between the V=O distances (V=O(6), 1.626(2) Å; V=O(7), 1.629(2) Å). These distances are similar to the V=O distances in (NMe$_4$)[VO$_2$(dipic-OH)].H$_2$O (1.6264(17), 1.6290(17) Å) and slightly longer than those observed in the parent compound Cs[VO$_2$(dipic)].H$_2$O (1.610(6)/1.615(6) Å).[100] Otherwise bond distances and angles in the coordination sphere of the V(V) metal center in Na[VO$_2$(dipic-OH)].2H$_2$O are similar to the corresponding parameters in (NMe$_4$)[VO$_2$(dipic-OH)].H$_2$O and Cs[VO$_2$(dipic)].H$_2$O.[100] The most significant difference among these structures arises from the formation of a polymeric structure as a result of the interactions of the sodium ion with coordinated dipic-OH^{2-} ligands.[99]

The reaction between [VO(dipic)(H$_2$O)$_2$].H$_2$O and creatinine resulted in the formation of a bis(oxo-bridged) binuclear vanadium(V) compound of stoichiometry [CH$_3$NHC(NH$_2$)$_2$]$_2$[V$_2$O$_4$(dipic)$_2$] (where [CH$_3$NHC(NH$_2$)$_2$]$^+$ = methyl gaunidinium).[80] An ORTEP drawing of the binuclear vanadium(V) complex with the atom numbering scheme is shown in Figure 11. The [VO$_2$(dipic)]⁻ ions are arranged in the lattice as centrosymmetric oxo-bridged binuclear complexes. The pair of vanadium(V) atoms in a dimer is in an edge sharing octahedral environment, with the dioxo vanadium(V) cation coordinated to a dipicolinate molecule acting as a tridentate ligand through one oxygen of each carboxylic group [V–O distances of 1.984(1) and 1.995(1) Å] and the heterocyclic nitrogen atom [d(V–N) = 2.097(2) Å].[80] The dipicolinate group defines an equatorial ligand plane [with a rms deviation of atoms from the least-squares plane of 0.025 Å] with the metal lying close to this plane [at 0.144(1) Å]. The bridging oxo ligand [d(V–O2) = 1.684(1) Å] is much closer to this plane [at 0.486(2) Å] than the terminal oxo atom [at 1.739(2) Å] which shows a slightly shorter V=O bond distance [d(V–O1) = 1.606(1) Å] and occupies an axial position. The O1–V–O2 angle is 105.36(7)°. The octahedral bonding structure around the metal is completed at the other axial position by the weak interaction with the bridging oxo ligand of the inversion related VO$_2^+$ group in the dimer [d(V–O2') = 2.370(1) Å].[80] The bonding structure around vanadium(V) agrees well with other binuclear dioxovanadium(V) complexes of tridentate ligands reported in the literature.[101] The above structural data was compared with that reported for the monomeric complex, Cs[VO$_2$(dipic)].H$_2$O.[100] Here the V=O double bond distances are equal to within experimental accuracy [1.610(6) and 1.615(6) Å], while in the dimeric complex, the V–O2 bond involving the bridging oxygen atom is 0.078(2) Å longer than the terminal V–O1 distance. This slight bond asymmetry within the dioxovanadium(V) ion is due to the formation of the weak intermolecular V•••O bond bridging the halves of the binuclear complex. Significant bond localization is also observed in the carboxylic groups of the dipicolinate ligand. In fact, the terminal C–O bond lengths are about 0.1 Å shorter than the C–O distances involving the coordinated-to-vanadium oxygen atoms. The C–N geometrical parameters of the methylguanidinium cation (MG) are in good accord with those found in related structures.[102-104] It has a singular planar CN$_3$ skeleton with a strong electronic delocalization that makes it able to participate in nets of H-bonding.

Figure 11. A diagram of the [V$_2$O$_4$(dipic)$_2$]$^{2-}$ anion. (Reproduced by permission from reference 80).

The $[VO_2(dipic)]^-$ monomeric units and the $[CH_3NHC(NH_2)_2]^+$ ions are arranged in the lattice along layers parallel to (101) crystallographic planes. Adjacent layers in the crystal are linked by the dimer-bridging bond. These layers are stabilized by a net of medium to strong N–H•••O bonds involving the ethylguanidinium NH and NH_2 groups and the oxo-bridge and the carboxylic oxygen atoms of the dipicolinate ligand [N•••O distances are in the range 2.858–2.964 Å and N–H•••O angles from 156.0° to 176.6°].[80] Each methylguanidinium has two N–H groups, from different N atoms, linked to one monomeric $[VO2(dipic)]^-$ unit through an oxo-bridge atom and the carboxylate oxygen coordinated to the metal. The structural feature of this pattern is similar to that described for the interaction of phosphate, sulfonate and nitrate with the guanidinium cation.[102-104] The other N–H groups of the same MG ion bond to adjacent monomer units through N–H•••O hydrogen bonds, forming a layer structure.

A number of 4-substituted, dipicolinatodioxovanadium(V) complexes and their hydroxylamido derivatives were synthesized and characterized by X-ray crystallography.[105] The $Na[VO_2(dipic\text{-}NH_2)].2H_2O$ complex (shown in Figure 12) is reported to crystallize as a salt without any required crystallographic symmetry. Selected bond distances and angles are provided in Table 4. The bond angles about the vanadium(V) suggested either a distorted square-pyramidal or trigonal-bipyramidal structure with the latter geometry being emphasized in Figure 12. The two largest angles about the V atom are O3-V1-O4) 149.99°. and O2-V1-N1 = 126.05°. It was reported that when these angles were used to define τ,[106] then one of the oxo ligands (O1) becomes the 'apical' ligand and τ = (149.99 - 126.05)/60) 0.399. Since τ is close to 0.5, the coordination geometry for this complex is neither trigonal-bipyramidal nor square-pyramidal.[105] A more detailed comparison of this structure to similar structures is shown in Table 2 and discussed below.[81, 99, 100]

In $Na[VO_2(dipic\text{-}NH_2)].2H_2O$, the six-coordinate sodium cation is bound to two symmetry-related O2 oxo ligand atoms (2.418(3), 2.459(3) Å). In addition, symmetry-related oxygen atoms from the C1 carboxylate group (Na-O3) 2.406(3) Å, Na-O5) 2.627(3) Å) and both of the lattice water molecules (Na-O7 = 2.405(4) Å, Na-O8 = 2.271(4) Å) are bound to sodium. A long and very weak seventh interaction is also present between sodium and the other oxo oxygen atom (Na-O1 = 3.009(3) Å). The amino substituent forms a weak hydrogen bond to a water molecule (N2•••O7 = 2.884(5) Å), as does one of the carboxylate oxygen atoms (O6•••O7) 2.774(4) Å).

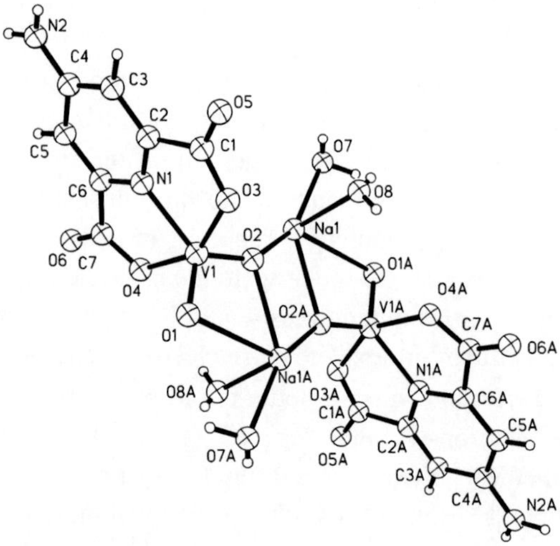

Figure 12. A diagram of Na[VO$_2$(dipic-NH$_2$)].2H$_2$O. (Reproduced by permission from reference 105)

The K[VO$_2$(dipic-NO$_2$)] complex (shown in Figure 13) also crystallizes without any required crystallographic symmetry; the pertinent bond angles and distances are listed in Table 3. The bond angles about the vanadium(V) suggested either a distorted square-pyramidal or trigonal-bipyramidal structure shown in Figure 13. The two largest angles about the V atom are O3-V-O4 = 148.81° and O1-V-N1 = 131.85°. Since the O2 atom is not involved in either of these two angles, it becomes the 'apical' ligand, and τ^{106} = (148.81 - 131.85)/60) 0.283. Since the τ value is closer to 0 than to 1, the coordination geometry for this complex is approaching square-pyramidal.[105] The bond distances and angles in these structures may be compared to the corresponding structural features of the Cs[VO$_2$(dipic)][100] and Na[VO$_2$(dipic-OH)],[81, 99] and Na[VO$_2$(dipic-NH$_2$)] (see Table 4). From the values in this table it is clear that the V=O and V-O bond lengths are virtually identical, showing that substitution in the 4-position of the pyridine ring does not affect those bonds. While a majority of the metric parameters in K[VO$_2$(dipic-NO$_2$)] are nearly identical to that of the amino and hydroxyl-substituted dipic^{2-} complexes, there are two notable differences. The first is the V-N$_{py}$ distance of 2.1019(17) Å, which is significantly longer than those bond lengths in complexes with electron donating substituents (Table 4). The other is the fact that the pyridine ring has a different orientation than in the other substituted dipicolinate complexes. [105] The O(1) atom is oriented in a slightly more trans fashion to the pyridine nitrogen (<N-V-O(1) = 131.85(7)°) than the O(2) atom which is in more of a *cis* orientation (<N-V-O(2) = 118.59(7)°).[105] This structural modification suggests that, with the appropriate ligand, it may be possible to introduce an additional ligand to form a six-coordinate complex. The N-V-O angles in the other substituted dipicolinate complexes are far more symmetric with respect to the dipicolinate ring, while the unsubstituted dipic structure displays somewhat more asymmetry (Table 4). In addition, the plane defined by the NO_2 group is twisted ~27° relative to the pyridine ring.

The V-N_{py} bond varies in length as would be expected from the electronic changes in the substituent group. The more electron donating substituent, NH_2, should make the pyridine N-atom a better σ and π donor and shorten the V-N_{py} bond, while the electron-withdrawing NO_2 group should have the opposite effect. Indeed the V-N_{py} bond length in [VO_2(dipic-NH_2)]⁻ is 2.050(3) Å compared to the V-N_{py} bond lengths in [VO_2(dipic-OH)]⁻ (2.077(4) and 2.0770(19) Å)24 and in [VO_2(dipic)]⁻ (2.089(6) Å);[100] the V-N bond length in [VO_2(dipic-NO_2)]⁻ is noticeably longer at 2.1019(17) Å. The eight-coordinate potassium ions knit the structure together tightly by binding to oxygen atoms from seven different symmetry-related complex anions. Five of the K-O bonds are markedly shorter than the other three. Three of these five shortest bonds to potassium involve oxo ligands O1 (2.7395(15) Å) and O2 (2.7495(15), 2.7681(15) Å). The other two short K-O interactions involve oxygen atoms from the nitro substituent (O7, 2.7139(15) Å) and one of the carboxylate groups (O5, 2.7838(16) Å). There is no hydrogen bonding in this structure due to a lack of protic donors.

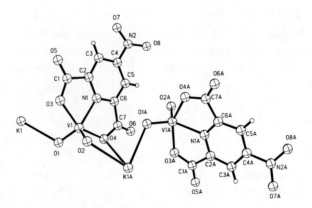

Figure 13. A diagram of K[VO_2(dipic-NO_2)]. (Reproduced by permission from reference 105).

Table 4. Selected bond lengths and angles for Na[VO$_2$(dipic-NH$_2$)].2H$_2$O, K[VO$_2$(dipic-NO$_2$)], and other five coordinate dipicolinato-vanadium(V) complexes.

Complex	V=O	V-N$_{py}$	V-O$_{carb}$	O$_{carb}$-V-O$_{carb}$	O$_{xo}$-V-N$_{py}$	O$_{carb}$-V-N$_{py}$	References
Na[VO$_2$(dipic-NH$_2$)].2H$_2$O	1.620(3) 1.627(3)	2.050(3)	1.990(3) 1.991(3)	149.99(12)	124.92(13) 126.05(13)	75.15(11) 74.85(11)	[105]
K[VO$_2$(dipic-NO$_2$)]	1.6253(15) 1.6293(14)	2.1019(17)	1.9910(14) 1.9953(15)	148.81(6)	118.59(7) 131.85(7)	74.75(6) 74.62(6)	[105]
Cs[VO$_2$(dipic)].H$_2$O	1.610(6) 1.615(6)	2.089(6)	2.001(5) 1.982(5)	149.4(2)	122.0(3) 128.2(3)	74.6(2) 75.9(2)	[100]
NMe$_4$[VO$_2$(dipic-OH)].H$_2$O	1.615(3) 1.626(3)	2.077(4)	1.008(4) 2.022(3)	148.88(14)	123.37(17) 125.92(18)	74.41(14) 74.47(7)	[99]
Na[VO$_2$(dipic-OH)].2H$_2$O	1.6264(17) 1.6290(17)	2.0770(19)	1.9945(16) 2.0011(16)	149.42(7)	124.48(8) 125.71(8)	74.96(7) 74.47(7)	[99]
K[VO$_2$(dipic-OH)].H$_2$O	1.606(5) 1.616(5)	2.089(6)	2.03395 1.990(5)	148.0(2)	123.1(3) 127.4(3)	73.3(2) 74.7(2)	[81]

The [VO(dipic)(Me$_2$-NO)(H$_2$O)].0.5H$_2$O complex (shown in Figure 14) crystallizes as discrete molecules without required crystallographic symmetry. Selected pertinent bond distances and angles are provided in Table 5 along with those of other crystallographically characterized, seven-coordinate vanadium(V)-dipic complexes.[89, 107-109] This complex contains seven coordinate vanadium in a pseudo, pentagonal-bipyramidal geometry. Alternatively, if one considers the hydroxylamido group to be a monodentate ligand, then the complex is a distorted, six-coordinate octahedral complex. The geometry of this complex is similar to that of the vanadium(V)-dipicolinato complex reported previously with the correspondingparent hydroxylamine, [VO(dipic)(H$_2$NO)(H$_2$O)].[89] Dimethylation of the hydroxylamine unit has no observable effects on the lengths of the V-O$_{H2O}$, V-O$_{R2NO}$, V-Npy, and the hydroxylamine N-O bonds. However, in the substituted hydroxylamine complex, the V-N$_{R2NO}$ bond length increased from 2.007(3) to 2.028(3) Å, respectively; such an increase would be anticipated based on the extra steric bulk of the two methyl groups. This steric bulk is also apparently sufficient to modify the coordination sphere of the vanadium resulting in a decrease of the V-O$_{carb}$ bond lengths, from 2.031(3) and 2.039(3) Å in **1a** to 2.008(3) and 2.026(3) Å in the substituted complex, **1c**.[105] Comparing the bond lengths of the hydroxylamido derivative complexes with those of the parent dipicolinato complexes (Tables 4 and 5) showed several differences. The V=O bonds are significantly shorter in the hydroxylamido complexes perhaps reflecting the need for additional electronic density in the more sterically crowded complexes. The V-N$_{py}$ bonds are shorter, whereas the V-O$_{carb}$ bonds are significantly longer in the hydroxylamido complexes compared to the parent complexes. Due to the fact that there is little change in the N-C-C and C-C-O angles, the shortening of the V-O$_{carb}$ bonds is primarily attributable to better overlap between the vanadium atom and the pyridine nitrogen donor.

Hydrogen bonding connects the complexes via interactions between the coordinated water molecule (O6) and carbonyl oxygen atoms O3 (2.817(5) Å) and O4 (2.764(5) Å). The disordered water molecule present in the lattice does not form any significant hydrogen bonds (perhaps accounting for its disorder).[105]

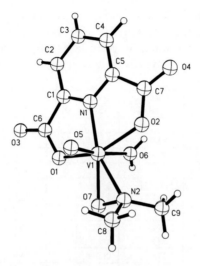

Figure 14. A diagram of [VO(dipic)(Me₂NO)(H₂O)]. (Reproduced by permission from reference 105).

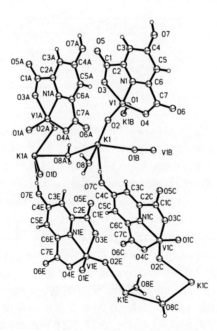

Figure 15. A diagram of K[VO₂(dipic-OH)].H₂O. (Reproduced by permission from reference 81).

Table 5. Selected bond length and angles for [VO(dipic)(Me₂NO)(H₂O)].0.5H₂O and other known seven-coordinate dipicolinato-vanadium(V) complexes with ternary peroxo or hydroxylamido ligands.

	[VO(dipic)(O₂)(H₂O)]	[VO(dipic)(O₂)(H₂O)]	[VO(dipic)(OOBuᵗ)(H₂O)]	[VO(dipic)(H₂NO)(H₂O)]	[VO(dipic)(Me₂NO)(H₂O)].0.5H₂O
Reference	107	109	108	89	105
V-O$_{oxo}$	1.579(2)	1.588(3)	1.574(3)	1.587(3)	1.588(3)
V-N$_{py}$	2.088(2)	2.080(3)	2.058(4)	2.064(3)	2.067(3)
V-N$_{R2NO}$				2.007(3)	2.028(3)
V-O$_{carb}$	2.053(2) 2.064(2)	2.041(3) 2.048(3)	1.996(3) 1.983(3)	2.031(3) 2.039(3)	2.026(3) 2.008(3)
V-O$_{R2NO}$				1.903(3)	1.902(3)
V-O$_{peroxo}$	1.870(2) 1.872(2)	1.869(3) 1.892(3)	1.872(3) 1.999(3)		
V-O$_{H2O}$	2.211(2)	2.235(3)	2.234(3)	2.240(3)	2.239(4)
N$_{R2NO}$-O$_{R2NO}$				1.371(4)	1.384(4)
O$_{peroxo}$-O$_{peroxo}$	1.441(3)	1.437(4)	1.436(5)		
O$_{carb}$-V-O$_{carb}$	147.2(1)	148.20(12)	149.4ᵃ	149.5(1)	148.44(12)
O$_{oxo}$-V-N$_{py}$	92.2(1)	95.53(16)	95.9(1)	97.2(1)	93.68(14)
O$_{oxo}$-V-O$_{carb}$	96.2(1) 94.9(1)	95.98(16) 94.38(16)	95.2(1) 94.1(1)	93.0(1) 98.9(1)	98.05(14) 95.08(13)
O$_{H2O}$-V-O$_{oxo}$	172.1(1)	169.84(16)	172.8(1)	172.0(1)	172.12(14)

ᵃ Not reported; calculated from the cif file.[105]

K[VO₂(dipic-OH)].H₂O was synthesized and characterized by X-ray crystallography and other physical techniques.[81] In Figure 15, the structure and the atom labeling scheme for K[VO₂(dipic-OH)].H₂O is shown. [VO₂(dipic-OH)]⁻ exists as a discrete mononuclear unit with the vanadium atom in a distorted trigonal bipyramidal coordination environment. The pyridine nitrogen atom (N1) and two oxygen atoms (O1, O2) of the VO₂ group coordinate to the vanadium center and occupy the distorted equatorial plane, while two carboxylate oxygen atoms occupy the axial positions. The vanadium atom is positioned 0.022(4) Å above the least squares equatorial plane through O1, O2, and N1. The chelation of the ligand decreases the angles around the vanadium center with carboxylate oxygen atoms and pyridine nitrogen atom (N1-V-O3 = 73.3(2)° and N1-V-O4 = 74.7(2)°) of the dipic-OH ligand with concomitant increase in other two angles (O3-V-O2 = 98.8(3)8 and O2-V-O4 = 100.4(3)°).[81] The constraints imposed by the chelation of the carboxylate oxygen atoms of the dipic-OH ligand at trans-positions result in a decrease of the O3-V-O4 bond angle from 180.08 to 148.0(2)°. As a result of the coordination of the carboxylate oxygen atoms of the dipic-OH ligand, the C1-C2-N1 and N1-C6-C7 angles on the pyridine ring are also decreased from 120.08 to 108.8(6)° and 110.8(6)°, respectively, with an increase of other corresponding bond angles.[81]

The VO_2 group is in the cis configuration, with an O1-V1-O2 angle of 109.5(3)° and with V-O1 and V-O2 distances of 1.606(5) and 1.616(5) Å, respectively. These V=O bonds are sufficiently short to imply double bonding with considerable π character. The V=O bond distance of $[VO_2(dipic-OH)]^-$ (1.606(5)/1.616(5) Å) is similar to the reported values for five-coordinate monooxovanadium(IV) complexes, including the Na^+ and NMe_4^+ salts of $[VO_2(dipic-OH)]^-$ (1.6264(17)/1.6290(17) and 1.615(3)/1.626(3) Å),[84] $[VO(bzac)_2]$ (1.612 (10) Å),[110] $[VO(acac-Et)_2]$ (1.605(2) Å),[111] and $[VO(acac-Me)_2]$ (1.592(2) Å).[111] V=O bond distances of the VO_2 group in six-coordinate vanadium(V) complexes is larger than that observed in this complex including $[VO_2(EDDA)]^-$ (1.632(1)/1.655(2) Å),[112] $[VO_2(EDTA)]^-$ (1.639(2)/1.657(1) Å),[113] and $[VO_2(pic)_2]^-$ (1.637(2)/1.638(2) Å).[114] The similarity of the V=O bond distance in five-coordinate V(IV) and V(V) complexes show that this bond distance is dependent upon the geometry around vanadium center and not on oxidation state. Long bonds extend from the vanadium atom to the carboxylate oxygen atoms (V-O3 = 2.033(5) and V-O4 = 1.990(5) Å) coordinated trans to each other at axial positions. These bond distances are in the range reported for other complexes including $[VO_2(pic)_2]^-$ (1.989(2) Å)[111] in which the carboxylate oxygen atoms are coordinated at axial positions. The nitrogen atom is coordinated in the distorted trigonal plane with V-N bond distance of 2.089(6) Å. This bond is, as expected, shorter than the bond distances reported for square pyramidal complexes in which nitrogen is coordinated in axial and/or equatorial plane[114, 115] and similar to the corresponding Na^+ and NMe_4^+ salts (2.0770(19) and 2.077(4)/2.070(4) Å).[84] The unit cell also contains one water molecule and a potassium ion. The water molecule resides between two monomer units making a hydrogen bond with one carboxylate oxygen atom of each of two monomer units (O5•••O8 = 2.893 Å, O4•••O8 = 2.921 Å) making a discrete dimer. The hydrogen bonding in this complex may be considered weak,[116] but the O8-H8b-O5 angle of 137.28 indicates that the hydrogen bonding is similar to those observed in organic structures.[117] The distances of the potassium ion from the oxygen atoms of the dioxo group (O1, O2), the oxygen atom of phenol OH (O7) and the oxygen atom of water molecule (O8) ranged from 2.664 to 2.878 Å which is normal for K$^+$•••O contacts.[118] However, the interactions of the potassium ion with O1 and O2 are not strong (2.793 and 2.878 Å) enough to have a significant affect on the V=O bond distance.[81]

K[V(C$_8$H$_3$NO$_6$)O$_2$].H$_2$O, was synthesized by reacting 4-carboxypyridine-2,6-dicarboxylic acid (contaminated as a potassium salt) with NH$_4$VO$_3$ in aqueous solution.[90] The complex, with a vanadium(V) metal center, is a distorted square-based pyramid (Figure 16). Its structure consists of chains of the anionic complexes in the direction of the b axis connected by potassium–oxygen interactions which range from 2.5981(18) to 3.0909(18) Å.[90] These chains are linked to each other by hydrogen bonding between the O atoms of the complex and the water molecules.[90] Selected bond lengths for K[V(C$_8$H$_3$NO$_6$)O$_2$].H$_2$O are shown in Table 6.

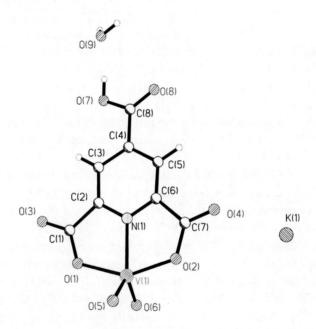

Figure 16. A diagram of K[V(C$_8$H$_3$NO$_6$)O$_2$].H$_2$O. (Reproduced by permission from reference 90).

Table 6. Selected bond lengths for K[V($C_8H_3NO_6$)O_2].H_2O.

Bond	Length/Å
V1-O6	1.6187(17)
V1-O5	1.6287(17)
V1-O2	1.9949(17)
V1-O1	2.0091(17)
V1-N1	2.086(2)

Payne et al. [119] recently reported the crystal structure of the [Cr(dipic)$_2$]$^-$ anion, with protonated 2,2'-dipyridylamine (Hdpa) as a counter ion. [(2-pyridyl)(1-hydro-2-pyridinium)amine][bis(2,6- pyridine dicar boxyla to) chromate (III)] trihydrate, 1, shows N_2O_4 coordination of the chromium (III) anion that is provided by two dianionic ligands, dipicolinate (Figure 17).[119] The distorted octahedral geometry of the chromium(III) metal center compares favorably in bond lengths and angles to that of the previously reported structure containing a rubidium cation.[120] Table 7 shows selected bond distances and angles for [Hdpa][Cr(dipic)$_2$]·3H_2O.

The [Hdpa]$^+$ cation (Figure 18) shows an isolated protonation at the N4 atoms with no positive residual electron density located near N5 in the final difference Fourier maps. No differences were observed in the bond distances of either of the formally pyridine and pyridinium rings. Bond localization is observed in the pyridinium-amine bond. A shortening of 0.037 Å is observed in the pyridinium-amine bond length while the pyridine-amine bond length compares favorably to a previously published structure of dpa.[121] In that structure, the dpa molecule crystallizes as a hydrogen bonded dimer in which the pyridine-amine bond length was found to be 1.380(4) Å. The rings of the cation in [(2-pyridyl) (1-hydro-2-pyridinium) amine] [bis (2, 6-pyridine dicarboxyla to)chromate(III)] trihydrate are twisted out of plane by 5.66(1)°. The rings of the [Hdpa]$^+$cations stack along the a axis. A packing diagram (Fig 19) of [(2-pyridyl) (1-hydro-2-pyridinium) amine] [bis (2, 6-pyridinedicarboxylato) chromate(III)] trihydrate is viewed approximately down the a axis. The water molecules O11 and O12 alternate to form approximately tilted square hydrogen bonded tetramers at the corners of the b/c cell edge along the a axis. These squares are then hydrogen bonded to the non-ligated carboxylate oxygen atoms O2 and O8 of four alternating anions. The remaining water molecule has O10 sitting at the center of a hydrogen bonded triangle formed with NH_3 of a cation and two oxygen atoms O5 (ligated) and O2 (non-ligated) of alternating anions.[119] The hydrogen bonding between O5–O10–O2 form a network between the anion layers.[119] The packing also consists of three π-π ring interactions of less than 3.8 Å.[122] These interactions are detailed in Table 7. There are two anion-anion interactions and one cation-cation interaction which are all related by an inversion center of each of the individual ring components.

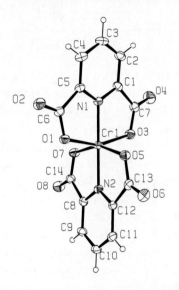

Figure 17. An ORTEP view of the anion of [(2-pyridyl)(1-hydro-2-pyridinium) amine] [bis (2,6-pyridinedicarboxylato)chromate(III)] trihydrate shown with 30 % probability ellipsoids and the atom numbering scheme. (Reproduced by permission from reference 119).

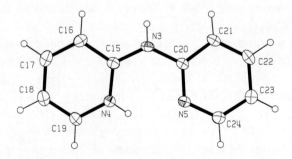

Figure 18. An ORTEP view of the cation of [(2-pyridyl)(1-hydro-2-pyridinium)amine][bis(2,6-pyridinedicarboxylato)chromate(III)] trihydrate shown with 25 % probability ellipsoids and the atom numbering scheme. (Reproduced by permission from reference 119).

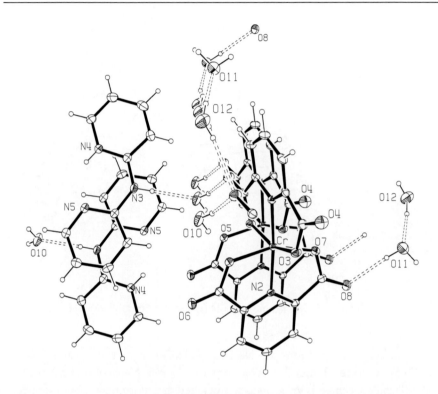

Figure 19. Selectively labeled ORTEP packing diagram of [(2-pyridyl)(1-hydro-2-pyridinium) amine] [bis(2,6-pyridinedicarboxylato)chromate(III)] trihydrate viewed approximately down the *a* axis. The thermal ellipsoids are drawn at the 20% probability level. (Reproduced by permission from reference 119).

Table 7. Selected Bond Distances (Å) and Angles (°) for [Hdpa][Cr(dipic)-$_2$]·3H$_2$O.

Cr-O3	1.9683(17)	O3-Cr-N1	79.30(7)	N2-Cr-O1	98.74(7)
Cr-N1	1.9706(19)	O3-Cr-N2	103.31(7)	O7-Cr-O1	95.77(7)
Cr-N2	1.9770(19)	N1-Cr-N2	175.35(8)	O3-Cr-O5	93.72(7)
Cr-O7	1.9834(17)	O3-Cr-O7	90.33(7)	N1-Cr-O5	97.79(7)
Cr-O1	1.9995(17)	N1-Cr-O7	105.21(7)	N2-Cr-O5	78.29(7)
Cr-O5	2.0071(16)	N2-Cr-O7	78.75(7)	O7-Cr-O5	157.00(7)
N3-C15	1.354(3)	O3-Cr-O1	157.88(7)	O1-Cr-O5	88.93(7)
N3-C20	1.391(3)	N1-Cr-O1	78.59(7)	C15-N3-C20	129.4(2)

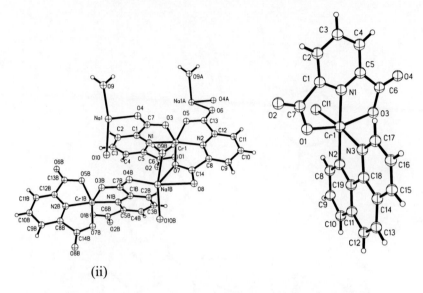

(ii)

Figure 20. Diagrams of Na[Cr(dipic)$_2$].2H$_2$O (i) and [Cr(dipic)(phen)Cl] (ii). (Reproduced by permission from reference 123).

The crystal structures of Na[Cr(dipic)$_2$].2H$_2$O and [Cr (dipic) (phen) Cl].0.5H$_2$O (see figure 20) were reported.[123] In Na[Cr(dipic)$_2$].2H$_2$O, the Cr(III) metal center is in a distorted octahedral environment, coordinated to two dipic^{2-} anions acting as tridentate ligands through its carboxylic oxygen atoms [Cr–O distances in the range from 1.985(5) to 1.998(4) Å] and the nitrogen atoms [d(Cr–N1) = 1.972(5) and d(Cr–N2) = 1.980(5) Å]. The dipic^{2-} ligands are nearly planar [rms deviation of atoms from the corresponding least squares planes less than of 0.07 Å] and perpendicular to each other [dihedral angle of 81.73(5)°]. The metal ion lies close onto the intersection of the coordination planes (along the N1•••N2 direction).[123]

The sodium ion is in a compact six fold coordination with four carboxylic oxygen atoms of neighboring dipic^{2-} groups [Na•••O distances in the range from 2.406(6) to 2.651(6) Å] and the two water molecules [d(Na•••O1w) = 2.329(6) Å and d(Na•••O2w) = 2.412(5)Å].[123]

For [Cr(dipic)(phen)Cl].0.5H$_2$O, the Cr(III) metal center is in a six-fold environment, coordinated to one dipic^{2-} ion defining a ligand equatorial plane [Cr–O distances of 1.975(3) and 1.999(3) Å and d(Cr–N) = 1.969(4) Å] and to a phen molecule acting as a bidentate ligand that bridges the fourth equatorial coordination site and one axial position through its nitrogen atoms [Cr–N distances of 2.075(4) and 2.081(3) Å, respectively].[123] The other axial position is occupied by a chlorine ion [d(Cr–Cl) = 2.290(1) Å]. The dipic^{2-} and phen species are nearly planar [rms deviation of atoms from the corresponding least-squares planes less than of 0.045 Å for both ligands] and close to mutual perpendicularity [angled at 86.60(7)° from each other]. As for the sodium salt, the Cr(III) ion nearly lies on the intersection of the coordination planes (along the N1•••N21 direction).[123]

The X-ray crystal structures of [Mn(dipic)(bpy)$_2$]·4.5H$_2$O and [Mn(chedam)(bpy)]·H$_2$O (chedam = chelidamic acid (4-hydroxypyridine-2,6-dicarboxylic acid) and bpy = 2,2'-bipyridine) were reported by Devereux *et al.*[124] The X-ray crystal structure of [Mn(dipic)(bpy)$_2$]·4.5H$_2$O is shown in Figure 21. The asymmetric unit consists of one tridentate dipicolinate ligand, two bidentate 2,2'-bipyridine molecules, four full occupancy water molecules and one half-occupancy water molecule. The [Mn(dipic)(bpy)$_2$] complex lies on a twofold axis, passing through Mn, N3 and C14 so that the by groups are *cis* to one another and the metal ion has irregular six-coordinate geometry. All the solvate water molecules and all the carboxylate oxygens of the dipic ligand are involved in hydrogen bonding. The ligands are involved in π-π stacking interactions with neighboring complex molecules.[124]

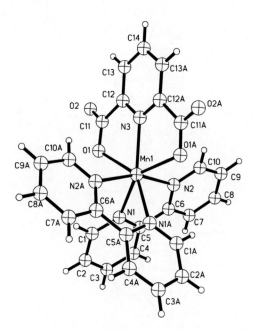

Figure 21. A diagram of [Mn(dipic)(bpy)]. (Reproduced by permission from reference 124).

The X-ray crystal structure of [Mn(chedam)(bpy)]·H_2O is shown in Figure 22. The structure contains monomeric [Mn(dicarboxylate)(bpy)H_2O] units in which the Mn(II) ions have very irregular six-coordinate geometry. This arrangement is imposed by the small bite angles of the diacid (O11- Mn-O15, 143.44(4)°) and bpy (72.19(5)8) ligands. Individual molecules are linked into chains running parallel to the c axis by hydrogen bonding involving the coordinated water molecule, the phenol group and the carboxylate groups and the chains are linked by π-π stacking of the bpy ligands parallel to the *b* axis.[124]

Iron, in various oxidations, is a very popular element which has been reported to be coordinated by dipicolinic acid.[19, 23-25, 125-132] One unique structure, [(dipic)$_2$(Hdipic)$_2$Fe$^{II}_3$(OH$_2$)$_4$], is reported in the literature.[23] In this unique structure, [(dipic)$_2$(Hdipic)$_2$Fe$^{II}_3$(OH$_2$)$_4$] crystallizes in the centrosymmetric space group *P2₁/n.* The asymmetric unit contains half a molecule located on a crystallographic inversion center on Fe(1) as shown on Figure 23 and a crystallization water molecule in general position. The structure consists of two [(dipic)(Hdipic)Fe^{II}] subunits sharing one oxygen atom each with a [O$_2$Fe^{II}(OH$_2$)$_4$] fragment.[23]

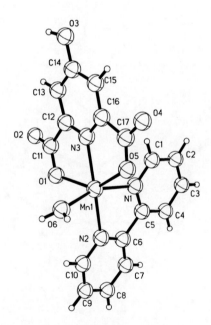

Figure 22. A diagram of [Mn(dipic-OH)(bpy)]. (Reproduced by permission from reference 124).

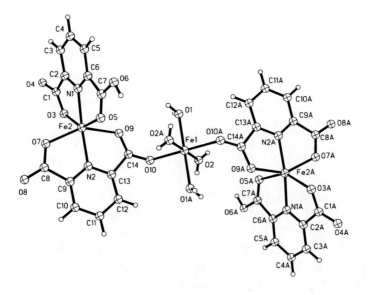

Figure 23. A diagram of [(dipic)$_2$(Hdipic)$_2$Fe$^{II}_3$(OH$_2$)$_4$]. (Reproduced by permission from reference 23).

The coordination sphere around Fe(1) is distorted with an elongation in the Fe(1)-O(110) direction as shown by the distances Fe(1)-O(110) = 2.161(1) Å, Fe(1)-O(120) = 2.085(2) Å, and Fe(1)-O(172) = 2.079(2) Å. It was reported that as observed for [(dipic)Fe^{II}(OH$_2$)$_3$], the shortest bond corresponds to a geometry in which the metal, the oxygen, and the two H atoms are coplanar (sums of the angles around the oxygen: 359° for O(120) and 335° for O(110)).[23] Here, O(110) is not involved in hydrogen bonding, whereas O(120) is linked to the oxygen atom O(111) of an another molecule with O(II)-O(120) = 2.71(3) Å, and O(ll)-H(127) = 1.98(4) Å. The geometry of the two [(dipic)(dipicH)Fe^{II}] subunits is very close to that of [(dipic)$_2$Fe^{II}]$^{2-}$. The acidic H atom is linked by hydrogen bonding to the water molecule with distances H(72)-O(100) = 1.64(4) Å, and O(72)-O(100) = 2.556(3) Å. The protonation of one of the carboxylate groups modifies (see structure II below) the bonding in the complex by the following: (i) a shortening of **a1**, here C(7)-O(71) (1.221(3) Å, to compare with 1.260(3), 1.279(3), and 1.275(3) Å for the similar C-O bonds); (ii) a loss of **b1** (here C(7)-O(72)) double bond character (1.302(3) Å to compare with C(1)-O(12) = 1.232(3) Å, C(11)-O(112) = 1.227(3) Å, and C(17)-O(172) = 1.248(3) Å); (iii) a lengthening of the **c1** bond, Fe(2)-O(71), 2.373(2) Å, whereas the other Fe-O bond **(d1)** ranges from 2.129(3) to 2.148(2) Å. It allows a relaxation of the compression of the Fe-N bond (2.102(2) Å), compared with Fe(2)-N(11) = 2.078(2) Å and less than 2.069(6) Å in [(dipic)$_2$Fe^{II}]$^{2-}$, and the loss of linearity of N-Fe-N' with an angle N(1)-Fe(2)-N(11) = 164.37(8)°.[23]

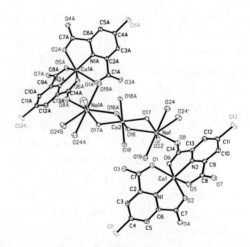

II

Reaction of H$_2$dipic and Mohr's salt at pH 6.25 gave the dinuclear species [(dipic)$_2$Fe^{II}(OH$_2$)$_5$], in which a [(dipic)$_2$Fe^{II}] moiety shares one oxygen atom with a [Fe^{II}(OH$_2$)$_5$O] fragment. [23] When the reaction was conducted at pH 5.85, the reaction gave crystals of (NH$_4$)$_2$[(dipic)$_4$(H$_2$dipic)$_2$Fe^{II}$_3$(OH$_2$)$_6$].4H$_2$O, which contains, in the unit cell, both the reactants [Fe^{II}(OH$_2$)$_6$]$^{2+}$ and H$_2$dipic and the product [(dipic)$_2$Fe^{II}]$^{2-}$.[23] At pH 5.65, the reaction yielded crystals of [(dipic)$_{10}$(dipicH)$_6$Fe^{II}$_{13}$(OH$_2$)$_{24}$].13H$_2$O, which contains, in the same unit cell, one trinuclear anionic complex ([[(dipic)$_2$Fe^{II}]-[Fe^{II}(OH$_2$)$_4$]-[Fe^{II}(dipic)$_2$])$^{2-}$, two doubly protonated cationic dinuclear complexes ((Hdipic)Fe^{II}(Hdipic)]-[Fe^{II}(OH$_2$)$_5$])$^{2+}$, two singly protonated cationic dinuclear complexes [(Hdipic)Fe^{II}(dipic)]-[Fe^{II}(OH$_2$)$_5$]$^+$, and two mononuclear anionic complexes [(dipic)$_2$Fe^{II}]$^{2-}$.[23]

Figure 24. A diagram of [Co$_3$Na$_2$ (C$_7$H$_2$ClNO$_4$)$_4$ (H$_2$O)$_{12}$] [Co (C$_7$H$_2$ClNO$_4$) (H$_2$O)$_3$]$_2$. 6H$_2$O. (Reproduced by permission from reference 41).

For many years now, cobalt complexes with dipicolinic acid have been of interest in coordination chemistry.[38, 56, 133-140] Recently, Moody et al.[141] reported the crystal structure of the very first cobalt(II) complex, $[Co_3Na_2(C_7H_2ClNO_4)_4(H_2O)_{12}][Co(C_7H_2ClNO_4)(H_2O)_3]_2.6H_2O$, with a substituted dipicolinate [4-chloropyridine-2,6-dicarboxylate = dipic-Cl] as ligand. $[Co_3Na_2(C_7H_2ClNO_4)_4(H_2O)_{12}][Co(C_7H_2ClNO_4)(H_2O)_3]_2.6H_2O$ (Figure 24), consists of a centrosymmetric dimer of $[Co^{II}(dipic-Cl)_2]^{2-}$ complex dianions [4-chloropyridine-2,6-dicarboxylate = dipic-Cl] bridged by an $[Na_2Co^{II}(H_2O)_{12}]^{4+}$ tetracationic cluster, two independent $[Co(dipic-Cl)(H_2O)_3]$ complexes, and six water molecules of crystallization.[141] The metals are all six-coordinate with distorted octahedral geometries. The $[Co^{II}(dipic-Cl)(H_2O)_3]$ complexes are neutral, with one tridentate ligand and three water molecules. The $[Co^{II}(dipic-Cl)_2]^{2-}$ complexes each have two tridentate ligands. The $[Na_2Co^{II}(H_2O)_{12}]^{4+}$ cluster has a central Co^{II} ion which is coordinated to six water molecules and lies on a crystallographic inversion center. Four of the water molecules bridge to two sodium ions, each of which have three other water molecules coordinated along with an O atom from the $[Co^{II}(dipic-Cl)_2]^{2-}$ complex. In the crystal structure, the various units are linked by O-H•••O hydrogen bonds, forming a three-dimensional network. Two water molecules are disordered equally over two positions.[141]

The synthesis and characterization of Co(II) and Co(III) dipicolinate ($dipic^{2-}$) complexes were reported by Yang et al.[56] X-ray crystallography was carried out on $[Co^{II}(H_2dipic)(dipic)].3H_2O$ and $[Co^{II}(dipic)(\mu-dipic)Co^{II}(H_2O)_5].2H_2O$. The molecular structure and atom labeling system for $[Co^{II}(H_2dipic)(dipic)].3H_2O$ is shown in Figure 25. Two nitrogen and four oxygen atoms are coordinated to the cobalt atom resulting in a distorted octahedron. Each of the two tridentate $dipic^{2-}$ ligands coordinates through two oxygen atoms and one nitrogen atom. One of the dipic groups is coordinated as $dipic^{2-}$ and the other as H_2dipic, resulting in this complex containing four of the different coordination modes (a-d) illustrated in Figure 3 above. The two Co-N bond distances are similar (2.017(3)-2.021(3) Å). Co-O bond distances range from 2.108(2) to 2.222(3) Å (average 2.166 Å).[56] The longest bond is observed in a type c coordination mode (Co-O(2)), and the shortest bond is observed in a type a coordination mode (Co-O(12)). The carboxylate group C(6)O(1)O(3) is coordinated to cobalt by the coordination mode shown in d, and the carboxylate C(7)O(2)O(4) is coordinated to the cobalt(II) by coordination mode c. The carboxylate groups C(16)-O(11)O(13) and C(17)O(12)O(14) are coordinated to the cobalt in coordination modes b and a, respectively. The identification of both a fully protonated neutral H_2dipic and a dianionic dipic^{2-} ligand in [Co^{II}(H$_2$dipic)(dipic)] distinguishes this structure from that of the nickel complex, [Ni(Hdipic)$_2$].3H$_2$O, which contains two monoprotonated Hdipic$^-$ ligands.[49, 50, 53] The structure of [Co^{II}(H$_2$dipic)(dipic)] exhibits some similarities with the structures of a silver(II) complex ([Ag(Hdipic)$_2$].H$_2$O)[22] and a copper(II) complex ([Cu(Hdipic)$_2$].3H$_2$O).[55]

The formula unit contains three water molecules, one of which is disordered in the crystal structure (O(40)). The other two water molecules form hydrogen bonds with the dianionic dipic ligand. Additional hydrogen bonding between water molecules O(20) and O(40) was found. The major component of O40 (i.e., O40A) is well defined, and hydrogen bonding distances are O(30)•••(4A) (1.54(5) Å), O(20)•••H1 (1.51(5) Å), O(13)•••H(20B) (1.48(5) Å), O(14)•••H(30A) (1.98(5) Å), and O(14)•••H(30B) (1.95(5) Å). Hydrogen bonds are formed between a H-donor and a H-acceptor, and those hydrogen bonds in which the proton originates in [Co^{II}(H$_2$dipic)(dipic)] are shorter than those in which [Co^{II}(H$_2$dipic)(dipic)] is the H-acceptor. An exception to this pattern is observed when the H-acceptor is a carboxylate group bound in coordination mode **b**. The two H$_2$O molecules (O(30) and O(20)) link two complexes through hydrogen bonds. The relevant distances are 1.98(5) Å (H(30A)•••O(14)), 1.54(5) Å (O(30)•••••••H(4A)), 1.48(5) Å (H(20B)•••••••O(13)) and 1.51(5) Å (O(20)•••••••H(1)). All hydrogen atoms except those of the disordered water oxygen atom O40 were found in the electron density map at the expected orientation.

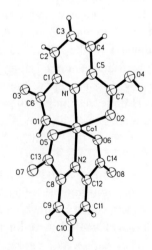

Figure 25. A diagram of [Co^{II}(H$_2$dipic)(dipic)].3H$_2$O. (Reproduced by permission from reference 56).

An ORTEP diagram of $[Co^{II}(dipic)(\mu\text{-}dipic)Co^{II}(H_2O)_5].2H_2O$ is shown in Figure 26.[56] The two $dipic^{2-}$ ligands are deprotonated in the complex. Both $dipic^{2-}$ ligands are coordinated in a tridentate manner to one cobalt atom; one of the two $dipic^{2-}$ groups also acts as a bridging ligand to the pentaaquo-Co(II) unit. This type of complex has previously been observed in $[Zn_2(H_2O)_5(dipic)_2].2H_2O$.[51] Both Co(II) ions exhibit distorted octahedral geometry. The Co-N bond distances are 2.026(2) and 2.033(1) Å, and the Co(1)-O bond distances range from 2.123(1) to 2.225(1) Å (average 2.182 Å). The carboxylate groups are bound in coordination modes **a** and **f**. The Co(2) atom is bound to six oxygen atoms, one from the carboxylate (Co(2)-O(14) 2.097(1) Å) and the other five from water molecules (Co-O from 2.060(2) to 2.180(1) Å, average Co(2)-O 2.097 Å for coordinated water). [56]

Strong hydrogen bonding exists in the crystal structure. Hydrogen bonding between coordinated water (O(23)) and the carboxylate oxygen atoms (O(2) and O(4)) of the $dipic^{2-}$ links the binuclear cobalt molecules to form a one-dimensional chain. The water molecule (O(30)) which hydrogen bonded to three complex molecules stabilizes the chain structure. Relevant distances are 1.81(3) Å (H(25B)••••••O(30)), 1.95(3) Å (H(30A)••••••O(3)), 2.08(3) Å (H(30B)••••••O(4)), 1.92(3) Å (H(23A)••••••O(4)), 2.10(3) Å (H(40B)••••••O(21)). The hydrogen bonds in this structure do not appear to be as strong as those in $[Co^{II}(H_2dipic)(dipic)].3H_2O$. [56]

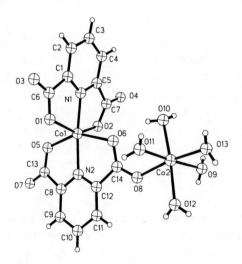

Figure 26. A diagram of $[Co^{II}(dipic)(\mu\text{-}dipic)Co^{II}(H_2O)_5].2H_2O$. (Reproduced by permission from reference 56).

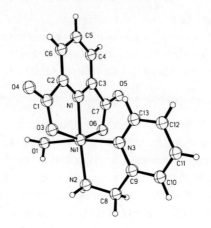

Figure 27. A diagram of [Ni(**1**)(dipic)(OH$_2$)]. (Reproduced by permission from reference 142)

Reaction of nickel(II) or copper(II) acetate with 2-(aminomethyl)pyridine **1** and the dipicolinate anion in aqueous methanol in a 1:1:1 molar ratio resulted in the formation of [Ni(**1**)(dipic)(OH$_2$)] (Figure 27) or [Cu(**1**)$_2$(CH$_3$OH)][Cu(dipic)$_2$] (Figure 28).[142] The nickel complex crystallizes as a highly hydrated structure with intricate hydrogen-bonding interactions, almost protein-like in character; that is elaborated by the complexes being ordered through π-stacking of the 2-(aminomethyl)pyridine ligands and hydrogen bonding in a type of a 'dimer tape' polymer arrangement. The asymmetric unit consists of two independent [Ni(**1**)(dipic)(OH$_2$)] molecules linked by close hydrogen bonds (Figure xxx), and ten water molecules. There are five fully occupied water molecules (O01, O02, O03, O05, and O06), one fully occupied water, but disordered over five positions (O010), and three half-occupied water molecules (O04, O07, and O08). The nickel centers feature bidentate coordination of **1** and tridentate coordination of dipic^{2-}, and with a coordinated water molecule. The structure exhibit a distorted octahedral *mer*-NiN$_3$O$_3$ geometry, with each axis involving a unique pair of donors (N$_{py}$ and N$_{amine}$; N$_{py}$ and O$_{water}$, O$_{COO}$ and O$_{COO}$).[142]

The two independent nickel complexes in the structure exhibit very similar distances and angles around the nickel metal centre. The two independent complexes are linked by hydrogen bonding interactions between the coordinated water of one nickel and a carboxylate oxygen of the other [O1•••O201 2.718(3) Å], reciprocated by the water and a carboxylate oxygen of the other center [O2•••O101 2.708(3) Å] were also reported.[142] Each coordinated water is also hydrogen bonded to the carboxylate oxygen of a symmetry related molecule [O1-O112 2.682(3) Å for Ni1 and O2•••O212 2.682(3) Å for Ni2]. Distortion in the octahedron around the nickel is also clear in figure 27. The planar tridentate chelation of dipic^{2-} leads to the O212-Ni2-O201 angle being reduced to 155.13(7)°, with intrachelate angles such as N205-Ni2-O212 at 77.72(8)° severely compressed compared with the relatively unstrained angle of the coordinated water N205-Ni2-O2 of 92.20(9)°. The other meridional angle involving **1** and the coordinated water, N220-Ni2-O2 of 172.42(9)°, also reflects some folding back, but not to the same extent as observed with coordination of dipic^{2-}. The Ni-N(pyridine) distance in the tridentate dipic^{2-} is significantly compressed [average 1.981(2) Å] compared with the distance of chelate **1** [average 2.073(2) Å]. The distances of Ni-O vary from an average of 2.134(2) Å for the carboxylate oxygens in dipic^{2-} to 2.085(2) Å for H$_2$O, despite the ionic nature of the former oxygen expecting a smaller distance. Both effects reflect the small bite angle of dipic^{2-} for the relatively large Ni(II) cation. The previously reported [Ni(Hdipic)$_2$].3H$_2$O complex exhibits an almost regular octahedral geometry,[49, 53] with the Ni-O distances in the range 2.10-2.21 Å, the longer distances being associated with the protonated acid group; the Ni-O(carbaoxylate) distances are comparable with the current Ni(II) complex, as are the Ni-N distances.[142]

The distortions due to the bidentate ligand **1** are reduced due to longer intra-ligand distances in the arm pendant to the pyridine ring compared to the case for dipic^{2-} and a longer preferred Ni-N(amine) distance compared to Ni-O(carboxylate); nevertheless, the N220-Ni2-N213 angle, for example, is compressed somewhat, at 80.96(9)°. Throughout the structure, planarity of the pyridine rings is maintained, albeit with some minor distortions reflected in ellipticity of probability ellipsoid for some ring atoms distant from the coordination sites.[142]

The copper(II) complex asymmetric unit contains two types of Cu(II) ions and three solvent methanols, consisting of independent cations [Cu(1)$_2$(CH$_3$OH)]$^{2+}$ and independent anions [Cu(dipic)$_2$]$^{2-}$ (Figure 28). The oxygen of the two methanols (O1 and O01) are coordinated strongly and weakly to Cu2. The other methanol (O02) is involved in receiving a hydrogen bond from the amine (N408) of one of the Cu1 ions and donating a hydrogen bond to the carboxylates (O101 and O201) of a Cu(II) ion. There are many other close contacts throughout the system, the hydrogen bonding along the direction of Jahn-Teller elongation in the two distinct Cu(II) ions link these and one of the unbound methanols in a polymer-like "ribbon" that apparently contributes significantly to the driving force for formation of the isolated solid.[142]

The crystal structure of $[Cu(dipic)_2]^{2-}$ has been described before, but with a different cation.[143] Bond distances and angle in the current complex are very similar to the reported complex. The complex is isolated as a neutral $[Cu(dipic)(H_2dipic)].xH_2O$ form, where two protonated carboxylate groups of one ligand occupy elongated apical positions of the Jahn-Teller distorted octahedron. Both monohydrate and trihydrate have been reported in the past,[144] with Cu-N distances in the range 1.901-1.907 (dianion) and 1.995-2.003 Å (diacid), Cu-OCO⁻ 2.008-2.063 Å and Cu-OCOH 2.302-2.465 Å. In the present structure, Cu-N distances vary from 1.896 Å, comparable to the dianion of the neutral complex, to a longer 1.9555 Å where the N is a component of the tridentate ligand with the Jahn-Teller elongated Cu-carboxylates, reflecting a compensation necessary to avoid serious angle strain in the carboxylate arms of the planar ligand. This Cu-N distance is still shorter than the reported when the Jahn-Teller elongated distances are also protonated. The Cu-OCO⁻ distances of 2.044 and 2.049 Å are comparable to those in the literature, whereas the Jahn-Teller elongated distances of 2.336 and 2.353 Å lie at the shorter end of the range reported for Jahn-Teller elongated and protonated carboxylates.[142]

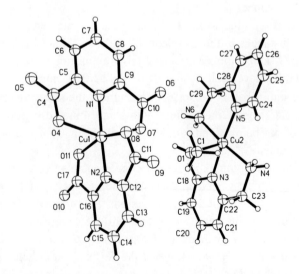

Figure 28. A diagram of $[Cu(1)_2(CH_3OH)][Cu(dipic)_2]$. (Reproduced by permission from reference 142).

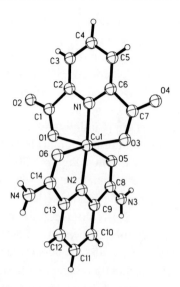

Figure 29. A diagram of [Cu(dipic)(pdcam)].2H$_2$O. (Reproduced by permission from reference 145).

The compound (2,6-pyridine-dicarboxamide) (2,6-pyridine-dicarboxylato) copper(II) dihydrate ([Cu(dipic)(pdcam)].2H$_2$O), was synthesized by the reaction of Cu$_2$(CO$_3$)(OH)$_2$, H$_2$dipic, and pdcam in an aqueous-ethanol medium.[145] [Cu(dipic)(pdcam)].2H$_2$O consists in elongated octahedral molecules where the copper(II) atom exhibits a coordination surrounding N$_2$O$_2$ + O$_2$, type 4 + 2 (Figure 29). The four closest donor atoms are the N-pyridine and two O-carboxylate donors of the tridentate dipic^{2-} ligand and the N-pyridine atom from the pdcam ligand.[145] Both primary amide O atoms from pdcam occupy the trans-apical positions of the Cu(II) coordination polyhedron, with bond lengths ~2.30 Å and define the lowest trans-angle of [Cu(dipic)(pdcam)].2H$_2$O (~150°). The trans-angle O(dipic)–Cu–O(dipic) also has a significantly low value (~160°). Both O–Cu–O trans angles of [Cu(dipic)(pdcam)].2H$_2$O reveal the rather rigid structures of such tridentate ligands, which are roughly planar (within 0.018 (dipic) or 0.079 (pdcam) Å. In contrast, the trans-angle N–Cu–N is very close to ~180° and the dihedral angle defined by the mean planes of dipic^{2-} and pdcam ligands is 88.4°, showing that they fall perpendicular. The lowest planarity of pdcam seems related to the implication of all N(amide)–H bonds in the 3-D H-bonding network of the crystal, with O carboxylate or water acceptor atoms. Stacking π,π-interactions between aromatic rings of the ligands are missing. On the other hand, all Cu–

N and Cu–O bond lengths of [Cu(dipic)(pdcam)].$2H_2O$ follow the ligand order dipic^{2-} < pdcam, probably because dipic^{2-} is a divalent anion whereas pdcam is a neutral ligand. The mer-NO_2(equatorial) conformation preferred by dipic^{2-} imposes to pdcam a mer-N(equatorial) + O_2(apical) conformation in the elongated octahedral Cu(II) coordination polyhedron, thus featuring its conformational flexibility.[145]

The molecular structures of [Cu(dipic)(4dmapy)] (where 4dmapy = 4-dimethylaminopyridine) and [Cu(dipic)(nmim)$(H_2O)_{0.5}$] (where nmim = N-methylimidazole) are shown in Figure 30.[146] The structure of [Cu(dipic)(4dmapy)] is square planar with the 4dmapy ligand coplanar with the CuN_2O_2 coordination plane, dihedral angle 1.3°. The Cu-O bond lengths of ~2 Å are normal, whereas the Cu-N distances are short. The four coordinate atoms lie slightly below the CoN_2O_2 best plane, N(1) 0.0174, N(2) 0.0184, O(1) 0.0040 and 0(3) 0.0042 Å, whilst the copper atom is above the plane by 0.0440 Å. The dimensions of the dipic ligand with bite angles of ~80° are in normal ranges. 4dmapy is mainly of the canonical form **lb**, as is evident by the short distances of C(10)-N(3), C(8)-C(9) and C(11)-C(12), and the essential co-planarity of the dimethylamine moiety and the pyridine ring. This is attributable to the delocalization of the amine lone pair with the pyridine nucleus.[146]

(1a)   **(1b)**

[Cu(dipic)(nmim)$(H_2O)_{0.5}$] consists of two distinct asymmetric copper units, a Cu(1) square planar unit [Cu(dipic)(nmim)] and a Cu(2) square pyramidal unit [Cu(dipic)(nmim)(H_2O)]. Both units contain a CuN_2O_2 square plane formed by the dipicolinate and the nmim ligand, where the Cu-N bond lengths are short and the Cu-O distances are normal. The dipic^{2-} bite angles are close to 80°. In the planar unit, the Cu(II) ion (0.0336 Å), O(1) (0.0034 Å] and O(3) (0.0031 Å) are slightly above the CuN_2O_2 best plane, while N(1) (-0.0208 Å) and N(2) (-0.0194 Å) lie slightly below. The dihedral angle between the imidazole nucleus and the CuN_2O_2 coordination plane is 6.8°. In the Cu(2) unit, N(4) (0.0688 Å), N(5) (0.0686 Å), O(5) (0.0031 Å), and O(7) (0.0044 Å) are slightly below the CuN_2O_2 best plane, and the copper(II) ion above (0.1449 Å). The dihedral angle between the imidazole nucleus and the CuN_2O_2 coordination plane is 13.9°. Hydrogen bonds were observed between the O(9) and O(4a) atoms (2.295 Å). The dimensions of the dipic^{2-} and nmim ligands are in the normal ranges.[146]

There are two complexation modes for dipicolinate copper(II) complexes; first, the dipic^{2-} binds strongly in the equatorial plane with a short Cu-N bond of ~1.9 Å and Cu-O bond of ~2.0 Å, as in [Cu(dipic)(H$_2$O)$_2$]$_2$,[54] secondly, the dipic^{2-} bonds perpendicularly to the equatorial plane with Cu-N of ~2.0 Å and Cu-O of ~2.4 Å, as in [Cu(dipic)(tpy)].[147] The [Cu(dipic)$_2$] moiety in [Cu$_2$(dipic)$_2$(bpy)$_2$].4H$_2$O[148] comprises both modes. The dipic ligands in both complexes belong to the first category, i.e., the dipic^{2-} ligands bind strongly in the equatorial coordination plane.[146]

The self-assembly of 4-hydroxypyridine-2,6-dicarboxylic acid (H$_3$CAM) and dipicolinic acid (H$_2$dipic) with Zn(II) salts under hydrothermal conditions gave two novel coordination polymers ([Zn(HCAM)]·H$_2$O)$_n$ and ([Zn(dipic)(H$_2$O)$_{1.5}$])$_n$.[149] Figure 31 shows a portion of an ORTEP diagram for ([Zn(dipic)(H$_2$O)$_{1.5}$])$_n$.[149]

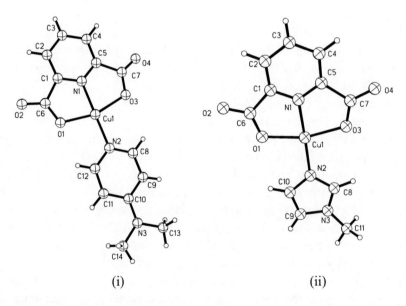

(i)   (ii)

Figure 30. Diagrams of [Cu(dipic)(4dmapy)] (i) and [Cu(dipic)(nmim)(H$_2$O)$_{0.5}$] (ii). (Reproduced by permission from reference 146).

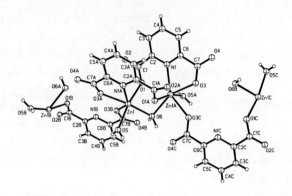

Figure 31. A diagram of ($[Zn(dipic)(H_2O)_{1.5}]$)$_n$. (Reproduced by permission from reference 149).

Three novel compounds were obtained as isomorphous ionic crystals containing $[M(dipicH_2)(OH_2)_3]^{2+}$ and $[Ce(dipic)_3]^{2-}$ ions.[150] The tri-capped trigonal prismatic Ce(IV) chelate has also been observed in Ce(IV)-alkaline earth-dipic complexes [151-154]. However, in all these cases some of the free oxygen atoms of the chelate coordinate to M(II) ions forming chains and networks. In the present cases there is no coordination link between the anions and cations. Instead, the free oxygen atoms form H-bonds with carboxylic groups and coordinated water molecules of the cation.[150]

All the three cations, $[M(dipicH_2)(OH_2)_3]^{2+}$, have very similar 4+2 coordination polyhedra. Figure 32 shows an ORTEP diagram for $[Zn(dipicH_2)(OH_2)_3][Ce(dipic)_3]$. The equatorial plane contains the nitrogen atom of dipicH$_2$ and the O atoms of the three water molecules, while the axial positions are occupied by the carbonyl oxygen atoms of dipicH$_2$. The major difference is that the axial elongation is significantly more in the case of the copper complex. The average equatorial and axial distances (Å) are 2.026(3), 2.175(3) (Ni); 1.984(2), 2.344(2) (Cu); 2.057(1), 2.236(1) (Zn). It would appear that the expected vibronic effects are superimposed on the steric requirements of the ligand to produce the observed static structure in the case of the copper complex. The average off-axis deviation (°) of the axial ligand atoms are 13.5 (Ni) and 15.1 (Cu, Zn). The equatorial atoms including the metal atoms are very nearly coplanar in all cases with a largest deviation of the atoms (Å) from the mean plane 0.038(3) (Ni), 0.062(2) (Cu) and 0.111(2) (Zn). The deviations amount to a very slight tetrahedral distortion with the following inter planar angles (°): 3.3(2) (Ni); 4.6(1) (Cu); 8.2(1) (Zn).[150]

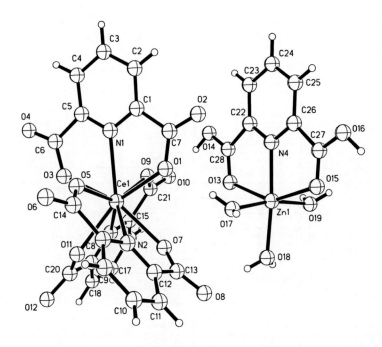

Figure 32. A diagram of [Zn(dipicH₂)(OH₂)₃][Ce(dipic)₃]. (Reproduced by permission from reference 150).

A very interesting report was published on the self-assembly of polyoxometallate clusters into a 3-D heterometallic framework via covalent bonding: synthesis, structure and characterization of $Na_4[Nd_8(dipic)_{12}(H_2O)_9][Mo_8O_{26}].8H_2O$,[155] but prior to that the crystal structures of [MoO₂(dipic)(L)] (L = DMF, DMSO, OPPh₃) were reported. [156] Single-crystal X-ray diffraction studies on [MoO₂(dipic)(L)] (L = DMF, DMSO, OPPh₃) revealed that the three complexes have a closely related structure, similar to that of [MoO₂(dipic)(DMF)] (see Figure 33).[156] The molybdenum atom adopts the expected distorted octahedral geometry, which is determined by the usual way of coordination for the dipicolinate ligand (planar tridentate) to the angular cis-MoO₂ moiety. Selected bond lengths and angles are listed in Table 8.[156] The bond lengths for Mo-O(5) and Mo-O(6) are not identical, the former being shorter. While the Mo-O(6) distance, involving the oxygen trans to N(1), is nearly identical for all the complexes, the Mo-O(5) distance increases with the basicity of the trans ligand.[156]

Table 8. Selected bond lengths (Å) and angles (°) for [MoO₂(dipic)(DMF)], [MoO₂(dipic)(DMSO)], and [MoO₂(dipic)(OPPh₃)].

	[MoO$_2$(dipic)(DMF)]	[MoO$_2$(dipic)(DMSO)]	[MoO$_2$(dipic)(OPPh$_3$)]
Mo(1)-O(5)	1.674(2)	1.696(2)	1.687(2)
Mo(1)-O(6)	1.702(2)	1.703(2)	1.700(2)
Mo(1)-O(1)	2.009(2)	2.008(2)	2.003(2)
Mo(1)-O(4)	2.010(2)	2.017(2)	2.000(2)
Mo(1)-N(1)	2.190(2)	2.198(2)	2.205(2)
Mo(1)-O(7)	2.308(2)	2.312(2)	2.247(2)
O(4)-C(7)	1.320(4)	1.335(3)	1.326(3)
O(1)-C(1)	1.325(4)	1.334(3)	1.325(3)
O(7)-X	1.233(4)	1.544(2)	1.502(2)
O(5)-Mo(1)-O(6)	105.5(1)	104.7(1)	104.6(1)
O(1)-Mo(1)-O(4)	144.68(9)	144.81(7)	145.43(8)
O(6)-Mo(1)-O(4)	106.5(1)	101.82(8)	103.86(9)
O(6)-Mo(1)-O(1)	101.8(1)	106.90(9)	104.73(8)
N(1)-Mo(1)-O(4)	72.48(8)	72.56(7)	72.95(8)
N(1)-Mo(1)-O(1)	72.79(8)	73.12(7)	73.03(7)
O(5)-Mo(1)-O(4)	95.8(1)	94.49(9)	72.95(8)
O(5)-Mo(1)-O(1)	96.6(1)	97.40(8)	95.79(9)
O(6)-Mo(1)-O(7)	82.3(1)	83.86(8)	85.04(8)
N(1)-Mo(1)-O(7)	71.76(8)	73.97(6)	71.31(7)
N(1)-Mo(1)-O(5)	100.5(1)	97.40(8)	99.06(9)
O(5)-Mo(1)-O(7)	172.0(1)	171.3798)	170.36(9)
N(1)-Mo(1)-O(6)	153.9(1)	157.58(8)	156.35(9)

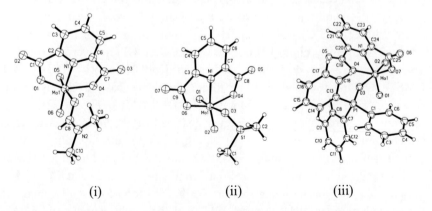

(i)   (ii)   (iii)

Figure 33. Diagrams of [MoO₂(dipic)(DMF)] (i), [MoO₂(dipic)(DMSO)] (ii), and [MoO₂(dipic)(OPPh₃)] (iii). (Reproduced by permission from reference 156)

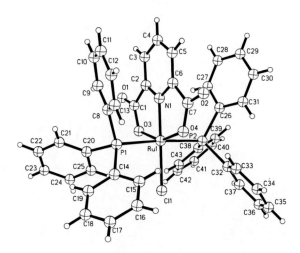

Figure 34. A diagram of [Ru(dipic)(PPh₃)₂Cl]. (Reproduced by permission from reference 157).

New hexa-coordinated Ru(III) complexes of the type [Ru(dipic)(EPh₃)₂X] (where X = Cl, Br; E = P, As) were synthesized by reacting dipicolinic acid with the appropriate starting complexes [RuX₃(EPh₃)₃].[157] The ligand behaves as tridentate dibasic chelate. A diagram of [Ru(dipic)(PPh₃)₂Cl] is shown in Figure 34. The N(1)–Ru(1)–Cl(1) bond angle is 178.95(9)° showing that Cl atom lies trans to ring nitrogen. The bite angles around Ru(III) are N(1)–Ru(1)–O(4) = 77.42(10)°; N(1)–Ru(1)–O(3) = 77.28(9)°; O(4)–Ru(1)–Cl(1) = 103.60(7)°, and O(3)–Ru(1)–Cl(1) = 101.70(6)°, summing up the in-plane angle to be exactly 360°. This shows the high planarity of the Cl and O, N, O donor atoms of dipicolinic acid. This was further supported by the other *cis* angles as reported.[157] The acid occupies the equatorial plane around the Ru(III) octahedron, along with Cl. The bond angle P(1)–Ru(1)– P(2) = 175.62(3)° shows that the two PPh₃ groups are *trans* to each other occupying the axial positions. The two Ru–P bonds are slightly bent away from the dipicolinic acid towards the Cl atom which is evident from the P(1)–Ru(1)–Cl(1) bond angle is 88.84° which is smaller than P(1)–Ru(1)–N(1) = 91.41° and P(2)–Ru(1)–N(1) = 92.51°.[157] The Ru–P, Ru–O, Ru–N and Ru–Cl bond lengths found in the complex agree well with that reported for similar ruthenium complexes.[157]

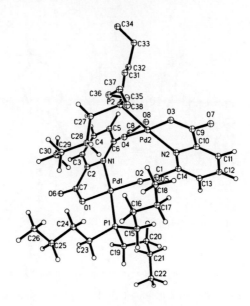

Figure 35. A diagram of [Pd(dipic)(PBu₃)]₂. (Reproduced by permission from reference 28).

The reactions of [Pd(acac)₂] or [Pd(OAc)₂]₃ with dipicolinic acid in acetonitrile produced [Pd(dipic)(NCMe)], which was used to synthesize [Pd(dipic)(PBu₃)]₂.[28] The X-ray structure and numbering scheme for [Pd(dipic)(PBu₃)]₂ are shown in Figure 35. The X-ray determination confirms the structure proposed on the basis of spectroscopic data.[28] The molecule is a dimer in which each dipic²⁻ ligand is coordinated to one palladium in a chelate manner through one carboxylate oxygen atom and the pyridine nitrogen, while the other carboxylate group, twisted out of the pyridine plane, is bonded to the second palladium atom of the dimer. The two bridges are arranged in a complementary head-to-tail manner. The resulting coordination around each palladium is distorted square planar with the remaining fourth coordination site filled by the phosphorus atom of the tributylphosphine.[28]

In related complexes the pyridine ring and the chelating carboxylate are nearly coplanar with the coordination plane.[26, 27, 158] However, in [Pd(dipic)PBu₃]₂ the pyridine ring is twisted by 20.0(2)° (for that on Pd(1)), or 12.2(2)° (for that on Pd(2)), with respect to the coordination planes. The deviation from coplanarity of the chelated carboxylate groups with respect to the pyridine ring is reflected in the torsion angles O(1)-(11)-C(12)-N(1) = -3(1)°, and O(2)-C(21)-C(22)-N(2) = -7(1)°. The bridging carboxylate groups are rotated with respect to the pyridine ring by 36.4(4) and 42.1(4)°. The C-O distances and angles for the chelate and bridged carboxylate groups are very similar since both carboxylates are coordinated. The $dipic^{2-}$ ligands are oriented in such a way that their aromatic rings are very nearly perpendicular to one another (dihedral angle 84.4(2)°), whereas the dihedral angle between the coordination planes of the two palladium atoms is 69.76(8)°. The phosphine seems to exert a marked influence on the *trans* Pd-N distances (2.140(5) and 2.139(6) Å). They are noticeably longer than the distances found in [Pd(dipic)Br]⁻ (2.034(8) Å)[159] or in dipicolinate complexes of platinum (1.88-2.03 Å)[26] and are similar to those found in the series of dimers $[Pd(\mu-\eta^2-pySN,S)Cl(PR_3)]_2$ (PR₃ = PMe₃, PMe₂Ph, and PMePh₂),[160] all of them having the PR₃ ligand in position *trans* to the N atom, which are in the range of 2.124(6)-2.137(9) Å.

The tendency of the dipicolinate complexes to dimerization is absent in the related complex [Pd(pdtc)PPh₃], as determined by ¹H NMR spectroscopy.[159] A plausible explanation for this different behavior is suggested by the structural features of both families. For [Pd(pdtc)Br]⁻ the chelating cycle is relatively free of strain (N-Pd-S) 86.44(6)°), whereas for [Pd(dipic)PBu₃]₂ the chelating carboxylate groups have a smaller bite angle (N(1)-Pd(1)-O(1) = 81.0(2)°; N(2)-Pd(2)-O(2) = 81.2(3)°). This is related to the shorter C-O distance (1.299(9) and 1.27(1) Å) compared to the C-S distance (1.711(9) Å, as well as by the longer Pd-N distance in [Pd(dipic)PBu₃]₂. Consequently the chelating cycles in the dipicolinate derivatives are more strained and have a higher tendency to relieve this strain by forming dimers (with only one strained cycle per palladium) rather than monomers (with two strained cycles per palladium), specially when the ancillary ligand induces a long Pd-N bond.[28]

A mixed Co-Ag complex with H₂dicpic, AgCo(dipic)₂, was synthesized under hydrothermal conditions.[161] X-ray structural analysis reveals that the asymmetric unit consists of one Ag(I) ion, one Co(III) ion and two $dipic^{2-}$ ligands. The complex crystallizes in the high symmetry space group $I4_1/a$. The Co(III) metal center has a distorted octahedral environment, with four *anti*-O(2) atoms of two $dipic^{2-}$ ligands in the equatorial plane and two N atoms in axial positions (figure 36). The two ²⁻ ligands are perpendicular to each other. The four *syn*-O(1) atoms of $dipic^{2-}$ are bonded at a distance of 2.4679(16) Å to Ag(I) to form an AgO₄ tetrahedron. All Co(III) and Ag(I) ions lie in a plane with a shortest Co···Ag contact of 4.795 Å. A search of the Cambridge Structural Database (version 5.26 [162]) revealed that the $[M(dipic)_2]^{n-}$ building block has been reported many times, but in those structures units are extended by interactions such as hydrogen bonding and π-stacking [57, 163, 164]. The present structure is the first example of 2-D sheet structure with 4^4 topology. Each carboxylate group associated with two [Co(dipic)$_2$]$^-$ ions links two Ag(I) ions to give a 4^4 tessellated 2-D net of Co$_2$Ag$_2$ groups with a Co···Ag separation of 4.975 Å.[161] Significant π-stacking involving pyridine rings is evident between the layers. The stacking propagates along the a and b axes in a decussated fashion (figure 4). This places pyridine rings of neighboring layers on top of one another with a separation of 3.10 Å (centroid–centroid = 3.50 Å, slippage angle = 27.5°). The rest of the layer contents stack by filling alternate bumps and hollows in adjacent layers.[161]

Although the π-stacking exhibits an offset face-to-face motif, the short separation distance indicates a strong π-π interaction.[161]

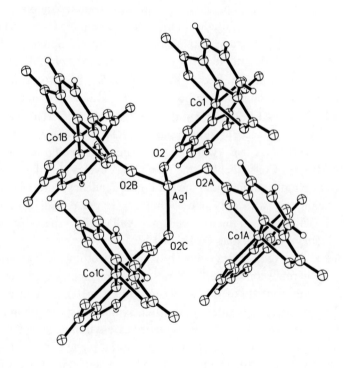

Figure 36. A diagram of the mixed-metal Co(III)-Ag(I) complex. (Reproduced by permission from reference 161)

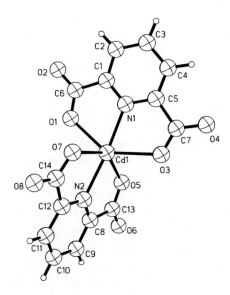

Figure 37. A diagram of the [Cd(dipic)$_2$]$^{2-}$ anion. (Reproduced by permission from reference 165).

A novel 1:2 proton transfer self-associated compound **LH2**, (GH$^+$)$_2$(dipic^{2-}), was synthesized from the reaction of dipicolinic acid and guanidine hydrochloride, (GH$^+$)(Cl$^-$).[165] The reaction of **LH$_2$** with cadmium(II) iodide in 2:1 molar ratio in water produced (GH)$_2$[Cd(dipic)$_2$]·2H$_2$O The molecular structure of the complex with atom numbering scheme and the crystal packing diagram are shown in Figure 37.[165] Important features of the crystal structure are the presence of an anionic complex [Cd(dipic)$_2$]$^{2-}$, cationic counter ion GH$^+$ and complexation of two pyridine dicarboxylates as tridentate ligands. The metal ion is hexacoordinated by two nitrogen atoms N(1), and N(9) and four oxygen atoms O(1), O(3), O(5) and O(7) of carboxylato groups of two dipic^{2-}. Fig. 2 shows that the Cd^{II} metal centre is located in the center of a distorted octahedral arrangement. The N(1)-Cd-N(9) angle shows deviation from linearity, 167.64(6)°. The C(6)-N(1)-Cd-O(7), C(6)-N(1)-Cd-O(5), C(10)-N(9)-Cd-O(3) and C(10)-N(9)-Cd-O(1) torsion angles are 101.77(14)°, -77.47(15)°, -81.78(14)°, and 102.33(14)°, respectively, indicating that two dianionic dipic^{2-} units are almost perpendicular to each other. Another characteristic solid state structural feature of this complex is dictated by the presence of guanidine fragment as a strong base that fully deprotonates pyridinedicarboxylic acid. This can lead to the formation of a Cd^{II} complex in which ion-pairing, metal-ligand coordination and intensive hydrogen-bonding play important roles in the construction of its three dimensional supramolecular network. It has been shown that 2,6-pyridinediamine (pyda) has key a role in the construction of a number of metal complexes of this proton transfer compound.[166-168] In the case of the Cd[II] complex, for instance, a centrosymmetric dinuclear Cd[II] complex containing pydaH[+] cation is prepared in which the two metal fragments are linked *via* the central four-membered Cd_2O_2 ring and each cadmium atom is in the center of a distorted pentagonal bipyramid.[168] As it is clear, the substitution of the pydaH[+] cation with GH[+] significantly changes the structure of the resulting Cd[II] complex. These data confirm the influence of the counter ion in the complexation as well as the ligation and stereochemistry of the complex.[165]

An important feature in both **LH₂** and $(GH)_2[Cd(dipic)_2]\cdot2H_2O$ complex is the presence of GH[+] cation with D_{3h} symmetry. The three C-N bond lengths in GH[+] are very close in values being 1.329(3) Å, 1.328(3) Å and 1.328(3) Å. These bond distances that are intermediate between C-N and C=N bonds indicate the equally distributed positive charge on three guanidinium nitrogen as expected. The angles N-C-N are all close to 120°. Each GH[+] cation adopts a planar structure and the multiple-electron delocalized π bond is thus formed. It is interesting to note that the C-N bond lengths of various types of guanidinium containing complexes reported to date vary in a range from 1.279 to 1.402 Å.[165] Extensive hydrogen bondings between carboxylate, GH[+], and water molecules throughout the lattice of Cd[II] complex play important roles in stabilizing the crystal. The range of D-H•••A angles and the H•••1468; A and D•••1468; A distances indicate the presence of strong hydrogen bonding in the Cd[II] complex lattice.[165]

A unique tin-containing complex was recently solved by X-ray crystallography.[169]   X-ray   diffraction   investigation   of $[(C_2H_5)_3NH][Me_3Sn(dipic)]$ (Figure 38) shows that it is not a infinite polymeric chain but an oligomer with three trimethyltin centers linked by two dipic[2-] ligands and bounded by two water molecules coordinating the terminal tin atoms.[169] In the case of $[(C_2H_5)_3NH][Me_3Sn(dipic)]$, both the trinuclear oligomeric anion and the counter cation are subject to crystallographically imposed twofold axial symmetry. Thus Sn(2) and C(11) of the anion and N(2) and C(16) of the cation are in special positions on a twofold axis, while all other atoms of the asymmetric unit, including the methylene group involving C(15) which is disordered over two symmetry related sites of equal occupancy, are in general positions. In $[(C_2H_5)_3NH][Me_3Sn(dipic)]$, each dianion bridges two Sn centers via only one O atom derived from the monodentate carboxylate moiety. Owing to the coordinated water molecules, the environments of the two terminal Sn atoms are different to another Sn atom. As a result of the bidentate mode of coordination of the dicarboxylic acid, each Sn center is five-coordinate and exists in trigonal bipyramidal geometry with the O (derived of water and carboxyl groups, respectively) atoms occupying the axial sites [Sn(2)–O(3) 2.266(4), Sn(2)–O(3^1^) (symmetry code: -x + 3/2, -y + 5/2, z) 2.266(4), Sn(1)–O(1) 2.179(3), Sn(1)–O(5) 2.396(4) Å and O(3)–Sn(2)–O(3^1^) 165.9(2)°, O(1)–Sn(1)–O(5) 177.38(14)°].[169]

The C–O bond distances [C(1)–O(2) 1.225(6) Å, C(7)–O(4) 1.217(7) Å] associated with the non-coordinating O atoms are significantly shorter than the coordinating C–O bond distances [C(1)–O(1) 1.278(6) Å, C(7)–O(3) 1.252(6) Å]. The intramolecular Sn(2)•••O(4) of 3.542 Å, is not indicative of bonding interactions between these atoms. Although not involved in coordination to tin, the O(4) atom form significant intermolecular contacts in the crystal lattice.[169]

Through the coordinated water molecule to the terminal tin atoms of the polymer, the polymers are associated with each other via hydrogen bonds to the pendant O atoms of carboxyl groups and the N atoms derived of the pyridine ring, so that a 2-D network is formed. The distances of hydrogen-bonding, O(5)–H•••O(2^1^), O(5)–H•••O(4^1^) and O(5)–H•••N(1^1^), separations are 2.759, 2.792, and 2.876 Å, respectively. This confirms that the water molecules play an important role in the stabilization of the polymer.[169]

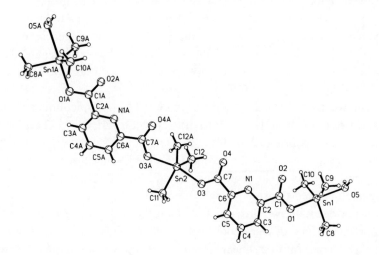

Figure 38. A diagram of the [Me₃Sn(dipic)]⁻ anion. (Reproduced by permission from reference 169)

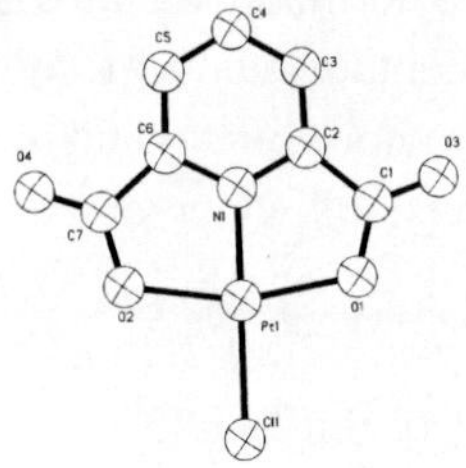

Figure 39. A diagram of the [Pt(dipic)Cl]⁻ anion. (Reproduced by permission from reference 26).

[Pt(dipic)Cl]⁻ was synthesized from dipicolinate salts and $[PtCl_4]^{2-}$.[26] Since the K^+ and $[Bu^n_4N]^+$ salts have identical spectral properties, the structure of the yellow form was characterized with the latter salt. Coexisting in the crystal of the yellow $[Bu^n_4N][Pt(dipic)Cl].0.5H_2O$ (see Figure 39) are three slightly different types of the complex anion. The exceptionally short Pt-N distance of 1.88-1.91 Å indicated partial multiplicity.[26] The M-N distances in the dipic chelates with first-row transition metals, whose atoms are much smaller than the Pt atom, span the range of 1.88-2.17 Å.[26] The O-Pt-N "bite" angles of 80.7-81.9° indicate strain in the chelate complex. The difference of ca. 0.10 Å between the two C-O distances in the same carboxylate group indicates considerable localization of the π electrons in the exocyclic group.[26]

There was a study where two compounds were reported to be coordination polymers,[170] but in one there are some Pb(II) cations bridged by carboxylate oxygen atoms which are only 4.355(4) Å apart, so they were regarded as lying in pairs, justifying an initial description in terms of a dimeric stoichiometric unit, $[Pb_2(dipic)_2(H_2dipic)_2(OH_2)_6]$ (Figure 40).[170]

The reaction of $(NH_4)_2Ce(NO_3)_6$ and $CaCl_2$ with dipicolinic acid resulted in formation of $[Ca(H_2dipic)(OH_2)_3][Ce(dipic)_3].5H_2O$.[171] Three tridentate dipic²⁻ ligands coordinate to Ce^{4+} in a slightly distorted tricapped trigonal prismatic mode (Figure 41). A tridentate dipicH₂ and three water molecules are bound to Ca^{2+}.[171] The protonated ligand is bound through the two carbonyl oxygen atoms and the ring nitrogen atom. Two of the dipic²⁻ ligands on Ce^{4+} act as a bridge with two different Ca^{2+} ions, thus completing a distorted square anti-prismatic type of eight-coordination around each Ca^{2+} ion). The resulting structure is an infinite linear chain made up of alternating Ce and Ca polyhedra). The Ce-Ca distance alternates between 6.409(2) and 6.831(2) Å along the chain. One of the lattice water molecules forms a strong H-bond (HO14•••OW4 = 1.47(9) Å, OW4•••HO14•••O14 = 164(7)°) with a COOH group of dipicH$_2$. While there is no previous report of a dipic complex of Ce(IV), Na$_3$[Ce(III)(dipic)$_3$].15H$_2$O was found[30] to be isomorphous with the Nd complex,[30] which has tricapped trigonal prismatic [Nd(dipic)$_3$]$^{3-}$ ions linked to form a three-dimensional network involving Na$^+$, carboxylate groups, and lattice water. Ca^{2+} forms a dinuclear complex [Ca(dipic)(OH$_2$)$_3$]$_2$,[33] in which Ca^{2+} is seven coordinate. A triiron linear chain is observed[23] in Fe$_3^{II}$ [(dipic)$_2$(Hdipic)$_2$(OH$_2$)$_4$], in which a central [Fe(OH)$_4$]$^{2+}$ unit is linked to two [Fe(dipicH)(dipic)]$^-$ units via carboxylate bridges.[171]

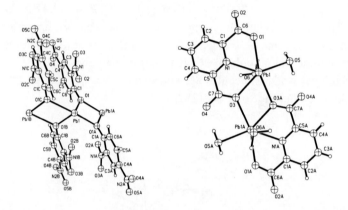

Figure 40. Diagrams of the simplest monomer and "dimer" present in [Pb$_2$(dipic)$_2$ (H$_2$dipic)$_2$(OH$_2$)$_6$]. (Reproduced by permission from reference 170)

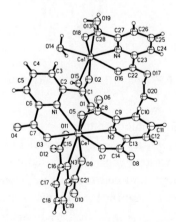

Figure 41. A diagram of [Ca(H$_2$dipic)(OH$_2$)$_3$][Ce(dipic)$_3$].5H$_2$O. (Reproduced by permission from reference 171).

The crystal structure of the newly reported $[Ce(dipic)_2(H_2O)_3].4H_2O$ reveals an unusual hydrogen-bonded water octamer.[172] The centrosymmetric octamer is built by bridging two water molecules to the chair form of a water hexamer. The structure, predicted to be unstable relative to other octameric structures, is stabilized by hydrogen bonding with the carboxylate groups lining the cavity in the host crystal.[172] The Ce(IV) is at the center of a distorted tricapped trigonal prismatic coordination polyhedron made up of three water molecules and two $dipic^{2-}$ ions coordinating in the tridentate chelating mode (Figure 42). The metal ligand bond distances (Å) are in the range Ce-$O_{carboxylate}$, 2.2273(1)-2.3692(1); Ce-O_{water}, 2.3790(2)-2.4728(2); Ce-N, 2.5171(2)-2.5218(2). The four lattice water molecules are assembled into a centrosymmetric octamer (Figure 2). The octamer has at its core a hexamer in the chair conformation with O•••O distances in the range 2.718(2)-2.788(2) Å. Two additional water molecules are attached at two diagonally opposite ends of the chair at a distance of 2.667(2) Å.[172]

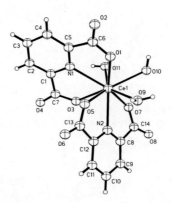

Figure 42. A diagram of $[Ce(dipic)_2(H_2O)_3].4H_2O$. (Reproduced by permission from reference 172).

X-ray crystallographic studies were carried out on $[Gua]_3[Ce(dipic)_3].3H_2O$ and $Na_3[Ce(dipic)_3].14H_2O$.[173] The use of guanidinium counter-cation considerably reduces the amount of water molecules per elementary unit cell from 14 in the case of sodium to three.[173] At the molecular scale, no significant differences were observed in the crystal structure of the $[Ce(dipic)_3]^{3-}$ moieties: each isomer (Λ or Δ) presents the classical features of the tris-dipicolinate lanthanate family with small deviation from the D3 symmetry and similar average Ce–O (2.519 and 2.515 Å) and

Ce–N (2.624 and 2.625 Å) distances for the Gua$^+$ and Na$^+$ complexes, respectively.[173] On the other hand, the crystal packing of both structures is very different as shown in Figure 43. In the case of the sodium derivative, the [Ce(dipic)$_3$]$^{3-}$ moieties are stacked in column, while the sodium cations are bridged by water molecules and carboxylate fragments in 1-D polymeric chains, the crystal cohesion being ensured by water molecule in a network of hydrogen bonds. On the contrary, in the case of the guanidinium compound, the crystal packing consists in alternated sheets composed by [Ce(dipic)$_3$]$^{3-}$ anions and guanidinium cations. The crystal cohesion is ensured by the guanidinum cations interconnecting two successive anionic sheets by hydrogen bonds with all the oxygen atoms of the dipicolinate fragments.[173]

The crystal structure of [Ho(Hdipic)(dipic)] was reported.[174] Dipic1 and dipic2 act as tridentate ligands towards the Ho^{3+} cation, with a pyridine nitrogen atom and two carboxylic oxygen atoms, (N(1), O(1), O(3)) and (N(2), O(5), O(7)), respectively (Figure 44). Moreover, dipic2 acts as a bis-monodentate ligand with two carboxylic oxygen atoms O(6) and O(8) (towards Ho^V and Ho^{IV}, respectively). Only the oxygen atoms O(2) and O(4) of dipic1 are not coordinated. One dipic anion is dianionic (dipic2), the other one (dipic1) must be monoprotonated to maintain electroneutrality. The proton was unambiguously located near O(1), from a difference Fourier synthesis. It is rather uncommon that the proton is located near an oxygen atom coordinated to the metal cation.[174]

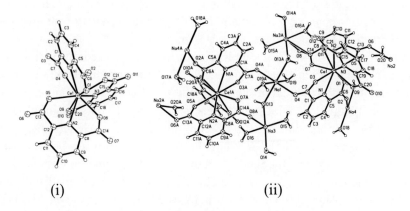

(i) (ii)

Figure 43. Diagrams of the [Ce(dipic)$_3$]$^{3-}$ anion (i) and Na$_3$[Ce(dipic)$_3$].14H$_2$O (ii). (Reproduced by permission from reference 173).

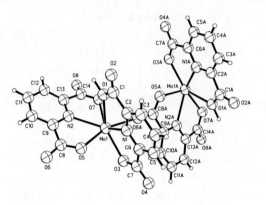

Figure 44. A diagram of [Ho(Hdipic)(dipic)]. (Reproduced by permission from reference 174).

Cs$^+$ and [Co(sar)]$^{3+}$ (where sar = 3,6,10,13,19-hexaazabicyclo [6.6.6] icosane] salts of the [Eu (dipic)$_3$]$^{3-}$ anion were chosen as materials suitable for initial structural studies because they were readily crystallized, contained an anion known usually to show strong visible luminescence.[175] The cesium and europium coordination environments are shown in Figure 45. The two cations can be considered to be bridged by the dipicolinate carboxylate groups. Two types of cesium ion, bridged by both carboxylate and water oxygen atoms, are found as part of a chain extending throughout the crystal. The chain can actually be considered as pairs of (bridged) eleven-coordinate cesium ions linked through bridging coordination to single eight coordinate cesium ions.[175] Six water molecules per formula unit of Cs$_3$[Eu(dipic)$_3$].9H$_2$O are involved in coordination to Cs$^+$, though all water molecules are involved in hydrogen bonding either to other water molecules or to the oxygen atoms of dipicolinate ligands. Although the space group is chiral and the crystal therefore optically active, the representation of the configuration of the complex anion is shown as Δ in Figure 1.[175] Metal-metal separations are of course significant with regard to electronic interactions affecting luminescence behavior; the shortest Cs•••Eu separation is 4.584 Å, while the shortest Eu•••Eu separation is 10.165 Å. Deviations of the [Eu(dipic)$_3$]$^{3-}$ anion from D$_3$ symmetry, despite the fact that the ion is situated on a twofold axis, are extremely small; the angles between the planes through the mid-point of the triangle O(11)-O(13)'-0(21)', Eu and each of the coordinated oxygens are 60.4, 59.7 and 59.9°, while the dihedral angle between planes O(11)-O(13)'-O(21)' and O(11)'-O(13)-O(21) is only 0.7°. The centroids of these planes are separated by 3.8 Å.[175]

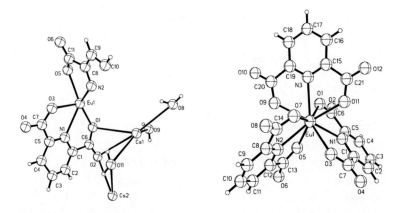

Figure 45. Diagrams of the [Eu(dipic)$_3$]$^{3-}$ anion. (Reproduced by permission from reference 175).

With regards to [UO$_2$(Hdipic)$_2$]·4H$_2$O (see Figure 46), the dipicolinate units function as tridentate ligands in both cases, the uranium coordination environment being close to pentagonal bipyramidal UNO$_6$ in the 1:1 complex and close to hexagonal bipyramidal UN$_2$O$_6$ in the 1:2 complex.[176] The lattice of [UO$_2$ (Hdipic)$_2$] ·4H$_2$O is similar to that of [NH (C$_2$H$_5$)$_3$]$_2$ [UO$_2$(dipic)$_2$]·2H$_2$O.[177]

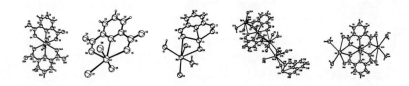

Figure 46. Diagrams of [UO$_2$(Hdipic)$_2$]·4H$_2$O. (Reproduced by permission from reference 176).

3.0. CONCLUSIONS

2,6-Pyridinedicarboxylic acid (dipicolinic acid) is definitely an excellent building block in coordination and supramolecular chemistry as outlined above.

REFERENCES

[1]   Froidevaux, P; Harrofield, JM; Sobolev, AN. *Inorg. Chem.*, 2000, 39, 4678-4687.

[2]   Haino, T; Matsumoto, Y; Fukazawa, Y. *J. Am. Chem. Soc.*, 2005, 127, 8936-8937.

[3]   Hunag, B; Prantil, MA; Gustafson, TL; Paquettte, J. R., *J. Am. Chem. Soc.*, 2003, 125, 14518-14530.

[4]   Muller, G; Schmidt, B; Jiricek, J; Hopfengadner, G; Riehl, J. P; Bunzli, JCG; Piguet, C. *J. Chem. Soc.*, *Dalton Trans.*, 2001, 2655-2662.

[5]   Storm, O; Luning, U. *Eur. J. Org. Chem.*, 2003, 3109-3116.

[6]   Devereux, M; McCann, M; Leon, V; McKee, V; Ball, R. J. *Polyhedron*, 2002, 21(11), 1063-1071.

[7]   Kapoor, R; Kataria, A; Pathak, A; Venugopalan, P; Hundal, G; Kapoor, P. *Polyhedron*, 2005, 24 (10), 1221-1231.

[8]   Kirillova, MV; Guedes da Silva, MFC; Kirillov, AM; Frausto da Silva, J.J.R; Pombeiro, A.J.L. *Inorg. Chim. Acta*, 2007, 360 (2), 506-512.

[9]   Ouali, N; Bocquet, B; Rigault, S; Morgantini, P.Y; Weber, J; Piguet, C. *Inorg. Chem.*, 2002, 41 (6), 1436-1445.

[10] Renaud, F; Piguet, C; Bernardinelli, G; Bunzli, J. C. G; Hopfgartner, G. *Chem.--Eur. J*, 1997, 3 (10), 1660-1667.

[11] Renaud, F; Piguet, C; Bernardinelli, G; Bunzli, J. C. G; Hopfgartner, G. *Chem.--Eur. J*, 1997, 3(10), 1646-1659.

[12] Tse, MK; Bhor, S; Klawonn, M; Anilkumar, G; Jiao, H; Doebler, C; Spannenberg, A; Maegerlein, W; Hugl, H; Beller, M. *Chem.--Eur. J*, 2006, 12(7), 1855-1874.

[13] Jackson, A; Davis, J; Pither, RJ; Rodger, A; Hannon, M. J. *Inorg. Chem.*, 2001, 40(16), 3964-3973.

[14] Takusagawa, F; Hirotsu, K; Shimada, A. *Bull. Chem. Soc. Jpn.* 1973, 46(7), 2020-2027.

[15] Moghimi, A; Sharif, M. A; Aghabozorg, H. *Acta Crystallgr*, 2004, E60(10), o1790-o1792.

[16] Smith, G; White, JM. *Aust. J. Chem.*, 2001, 54, 97-100.

[17] Ramsay, W., *Jahresber. Fortschr. Chem.*, 1877, 437.

[18] Udo, S. *Nippon Nogei Kagaku Kaishi*, 1936, 12, 386-394.

[19] Laine, P; Gourdon, A; Launay, JP. *Inorg. Chem.*, 1995, 34(21), 5129-5137.

[20] Ventur, D; Wieghardt, K; Weiss, J. *Z. Anorg. Allg. Chem.*, 1985, 524, 40-50.

[21] Drew, MGB; Fowles, GWA; Matthews, RW; Walton, RA. *J. Am. Chem., Soc*, 1969, 91(27), 7769-7771.

[22] Drew, MGB; Matthews, RW; Walton, RA. *J. Chem. Soc. A*, 1970, (9) 1405-1410.

[23] Laine, P; Gourdon, A; Launay, JP. *Inorg. Chem.*, 1995, 34(21), 5138-5149.

[24] Laine, P; Gourdon, A; Launay, JP. *Inorg. Chem.*, 1995, 34(21), 5156-5165.

[25] Laine, P; Gourdon, A; Launay, JP; Tuchagues, JP. *Inorg. Chem.*, 1995, 34(21), 5150-5155.

[26] Zhou, XY; Kostic, NM. *Inorg. Chem.*, 1988, 27(24), 4402-4408.

[27] Chessa, G; Marangoni, G; Pitteri, B; Bertolasi, V; Gilli, G; Ferretti, V. *Inorg. Chim. Acta*, 1991, 185(2), 201-210.

[28] Espinet, P; Miguel, JA; Garcia-Granda, S; Miguel, D. *Inorg. Chem.*, 1996, 35(8), 2287-2291.

[29] Albertsson, J. *Acta Chem. Scand*, 1970, 24(4), 1213-1229.

[30] Albertsson, J. *Acta Chem. Scand*, 1972, 26(3), 1023-1044.

[31] Albertsson, J. *Acta Chem. Scand*, 1972, 26(3), 985-1004.

[32] Palmer, KJ; Wong, RY; Lewis, JC. *Acta Crystallogr., Sect. B*, 1972, 28, (Pt. 1), 223-228.

[33] Strahs, G; Dickerson, RE. *Acta Crystallogr., Sect. B*, 1968, 24(4), 571-578.

[34] Groves, JT; Kady, I. O. *Inorg. Chem.*, 1993, 32(18), 3868-3872.

[35] Harrington, PC; Wilkins, RG. *J. Inorg. Biochem*, 1980, 12(2), 107-118.

[36] Holder, AA; Brown, RFG; Marshall, SC; Payne, VCR; Cozier, MD; Alleyne, WA., Jr; Bovell, CO. *Transition Met. Chem.*, 2000, 25(5), 605-611.

[37] Mauk, AG; Bordignon, E; Gray, HB. *J. Am. Chem. Soc.*, 1982, 104(26), 7654-7657.

[38] Mauk, AG; Coyle, CL; Bordignon, E; Gray, HB. *J. Am. Chem. Soc.*, 1979, 101(17), 5054-5056.

[39] Varey, JE; Lamprecht, GJ; Fedin, VP; Holder, A; Clegg, W; Elsegood, MRJ; Sykes, AG. *Inorg. Chem.*, 1996, 35(19), 5525-5530.

[40] Balavoine, G; Barton, DHR; Gref, A; Lellouche, I. *Tetrahedron*, 1992, 48(10), 1883-1894.

[41] Cofre, P; Richert, SA; Sobkowiak, A; Sawyer, DT. *Inorg. Chem.*, 1990, 29(14), 2645-2651.

[42] Dalton, H. *Adv. Applied Microbiol*, 1980, 26, 71-87.

[43] Ericson, A; Hedman, B; Hodgson, KO; Green, J; Dalton, H; Bentsen, J.

G; Beer, RH; Lippard, SJ. *J. Am. Chem. Soc.*, 1988, 110(7), 2330-2332.

[44] Fox, BG; Surerus, KK; Munck, E; Lipscomb, JD. *J. Biol. Chem.*, 1988, 263(22), 10553-10556.

[45] Sheu, C; Sawyer, DT. *J. Am. Chem. Soc.*, 1990, 112(22), 8212-8214.

[46] Sheu, C; Sobkowiak, A; Jeon, S; Sawyer, DT. *J. Am. Chem. Soc.*, 1990, 112(2), 879-881.

[47] Sugimoto, H; Tung, HC; Sawyer, DT. *J. Am. Chem. Soc*, 1988, 110(8), 2465-2470.

[48] Tung, HC; Kang, C; Sawyer, DT. *J. Am. Chem. Soc.*, 1992, 114(9), 3445-3455.

[49] Chiesi Villa, A; Guastini, C; Musatti, A; Nardelli, M. *Gazz. Chim. Ital.*, 1972, 102(3), 226-233.

[50] Gaw, H; Robinson, WR; Walton, RA. *Inorg. Nucl. Chem. Lett*, 1971, 7(8), 695-699.

[51] Hakansson, K; Lindahl, M; Svensson, G; Albertsson, J. *Acta Chem. Scand*, 1993, 47, 449-455.

[52] Okabe, N; Oya, N. *Acta Crystallogr., Sect. C: Cryst. Struct. Commun*, 2000, C56(3), 305-307.

[53] Quaglieri, P; Loiseleur, H; Thomas, G. *Acta Crystallogr., Sect. B*, 1972, 28(Pt. 8), 2583-90.

[54] Biagini Cingi, M; Chiesi Villa, A; Guastini, C; Nardelli, M. *Gazz. Chim. Ital*, 1971, 101(11), 825-832.

[55] Biagini Cingi, M; Chiesi Villa, A; Guastini, C; Nardelli, M. *Gazz. Chim. Ital*, 1972, 102(11), 1026-1033.

[56] Yang, L; Crans, DC; Miller, SM; la Cour, A; Anderson, OP; Kaszynski, PM; Godzala, ME; Austin, LD; Willsky, GR. *Inorg. Chem.*, 2002, 41(19), 4859-4871.

[57] MacDonald, JC; Dorrestein, PC; Pilley, MM; Foote, MM; Lundburg, JL; Henning, RW; Schultz, AJ; Manson, JL. *J. Am. Chem. Soc.*, 2000, 122(47), 11692-11702.

[58] Browning, K; Abboud, KA; Palenik, GJ. *J. Chem. Crystallogr*, 1995, 25(12), 851-855.

[59] Carmona, P. *Spectrochim. Acta Part A: Molecular Spectroscopy*, 1980, 36(7), 705-712.

[60] Starosta, W; Ptasiewicz-Bak, H; Leciejewicz, J. *J. Coord. Chem.*, 2002, 55(8), 873-881.

[61] Soleimannejad, J; Aghabozorg, H; Hooshmand, S; Adams, H. *Acta Crystallogr., Section E*, 2007, E63, (12), m3089-m3090, m3089/1-m3089/14.

[62] Gaetjens, J; Meier, B; Adachi, Y; Sakurai, H; Rehder, D. *Eur. J. Inorg. Chem.*, 2006, (18), 3575-3585.

[63] Manohar, H; Schwarzenbach, D. *Helv. Chim. Acta*, 1974, 57(4), 1086-1095.

[64] Leik, R; Zsolnai, L; Huttner, G; Neuse, EW; Brintzinger, HH. *J. Organomet. Chem.*, 1986, 312(2), 177-182.

[65] Anderson, SJ; Brown, DS; Norbury, AH., *J. Chem. Soc., Chem. Commun.*, 1974, 996.

[66] Sanner, RD; Dugga, DM; Mckenzie, TC; Marsh, RE; Bercaw, JE. *J. Am. Chem. Soc.*, 1976, 98, 8358-8365.

[67] Zeinstra, JD; Teuben, JH; Jellinek, F. *J. Organomet. Chem.* 1979, 170, 39-50.

[68] Besancon, J; Top, S; Tirouflet, J; Dusausoy, Y; Lecomte, C; Protas, J. *J. Organomet. Chem.*, 1977, 127, 153-168.

[69] Curtis, MD; Thanedar, S; Butler, WM. *Organometallics*, 1984, 3, 1855-1859.

[70] Huffman, JC; Moloy, kG; Marsella, JA; Caulton, KG. *J. Am. Chem. Soc*, 1980, 102, 3009-3014.

[71] Silver, ME; Eisenstein, O; Fay, RC. *Inorg. Chem.*, 1983, 22, 759-770.

[72] Thewalt, U; Lasser, W. *J. Organomet. Chem.*, 1984, 276, 341-347.

[73] Schwarzenbach, D. *Helv. Chim. Acta*, 1972, 55(8), 2990-3004.

[74] Smee, JJ; Epps, JA; Teissedre, G; Maes, M; Harding, N; Yang, L; Baruah, B; Miller, SM; Anderson, OP; Willsky, GR; Crans, DC. *Inorg. Chem.*, 2007, 46(23), 9827-9840.

[75] Smee, JJ; Epps, JA; Ooms, K; Bolte, SE; Polenova, T; Baruah, B; Yang, L; Ding, W; Li, M; Willsky, GR; Cour, Al; Anderson, OP; Crans, DC. *J. Inorg. Biochem*, 2009, 103(4), 575-584.

[76] Cocco, MT; Onnis, V; Ponticelli, G; Meier, B; Rehder, D; Garribba, E; Micera, G. *J. Inorg. Biochem*, 2007, 101(1), 19-29.

[77] Parajón-Costa, BS; Piro, OE; Pis-Diez, R; Castellano, EE; González-Baró, AC. *Polyhedron*, 2006, 25(15), 2920-2928.

[78] Kavitha, SJ; Panchanatheswaran, K; Elsegood, MRJ; Dale, SH. *Inorg. Chim. Acta*, 2006, 359(4), 1314-1320.

[79] Yin-Zhuang, Z; Jinli, L. *Inorg. Chem. Commun*, 2009, 12(3), 243-245.

[80] Gonzalez-Baró, AC; Castellano, EE; Piro, OE; Parajón-Costa, BS. *Polyhedron*, 2005, 24(1), 49-55.

[81] Crans, DC; Mahroof-Tahir, M; Johnson, MD; Wilkins, PC; Yang, L; Robbins, K; Johnson, A; Alfano, JA; Godzala, ME; Austin, LT; Willsky, GR. *Inorg. Chim. Acta*, 2003, 356, 365-378.

[82]   Bersted, BH; Belford, RL; Paul, IC. *Inorg. Chem.*, 1968, 7(8), 1557-1562.

[83]   Chatterjee, M; Ghosh, S; Wu, BM; Mak, TCW. *Polyhedron*, 1998, 17(8), 1369-1374.

[84]   Crans, DC; Yang, L; Jakusch, T; Kiss, T. *Inorg. Chem.*, 2000, 39(20), 4409-4416.

[85]   Xing, YH; Aoki, K; Bai, FY. *J. Coord. Chem.*, 2004, 57(2), 157-165.

[86]   Bersted, BH. Crystal, molecular, and electronic structure of vanadyl(IV) 2,6-lutidinate tetrahydrate. 1969, 94 CAN 74:104184 AN 1971:104184.

[87]   Hartkamp, H. *Angew. Chem.*, 1959, 71, 553.

[88]   Thompson, KH; Orvig, C. *Coord. Chem. Rev.*, 2001, 219-221, 1033-1053.

[89]   Nuber, B; Weiss, J. *Acta Crystallogr., Sect. B*, 1981, B37(4), 947-948.

[90]   Holder, AA; VanDerveer, D. *Acta Crystallogr., Sect. E: Struct. Rep. Online*, 2007, E63(8), m2051-m2052.

[91]   Casny, M; Rehder, D. *Chem. Commun.*, 2001, (10), 921-922.

[92]   Shaver, A; Ng, JB; Hall, DA; Lum, BS; Posner, BI. *Inorg. Chem.*, 1993, 32(14), 3109-3113.

[93]   Shaver, A; Hall, DA; Ng, JB; Lebuis, AM; Hynes, RC; Posner, BI. *Inorg. Chim. Acta*, 1995, 229(1-2), 253-260.

[94]   Kiss, E; Benyei, A; Kiss, T. *Polyhedron*, 2003, 22(1), 27-33.

[95]   Chatterjee, M; Ghosh, S; Nandi, AK. *Polyhedron*, 1997, 16(17), 2917-2923.

[96]   Farrugia, LJ. *J. Appl. Crystallogr*, 1997, 30(5, Pt. 1), 565.

[97]   Cornman, CR; Kampf, J; Lah, MS; Pecoraro, VL. *Inorg. Chem.*, 1992, 31(11), 2035-2043.

[98]   Cotton, FA; Czuchajowska, J; Feng, X. *Inorg. Chem.*, 1991, 30(2), 349-353.

[99]   Yang, L; la Cour, A; Anderson, OP; Crans, DC. *Inorg. Chem.*, 2002, 41(24), 6322-6331.

[100] Nuber, B; Weiss, J; Wieghardt, K. *Z. Naturforsch., B: Anorg. Chem., Org. Chem*, 1978, 33B(3), 265-267.

[101] Li, X; Lah, MS; Pecoraro, VL. *Inorg. Chem.*, 1988, 27(25), 4657-4664.

[102] Cotton, FA; Day, VW; Hazen, EE., Jr; Larsen, S. *J. Am. Chem. Soc.*, 1973, 95(15), 4834-4840.

[103] Katrusiak, A; Szafranski, M. *J. Mol. Struct*, 1996, 378(3), 205-223.

[104] Russell, VA; Etter, MC; Ward, MD. *J. Am. Chem. Soc.*, 1994, 116(5), 1941-1952.

[105] Smee Jason, J; Epps Jason, A; Teissedre, G; Maes, M; Harding, N;

Yang, L; Baruah, B; Miller Susie, M; Anderson Oren, P; Willsky Gail, R; Crans Debbie, C. *Inorg. Chem.*, 2007, 46(23), 9827-9840.

[106] Addison, AW; Rao, TN; Reedijk, J; Van Rijn, J; Verschoor, GC. *J. Chem. Soc., Dalton Trans.*, 1984, (7), 1349-1356.

[107] Drew, RE; Einstein, FWB. *Inorg. Chem.*, 1973, 12(4), 829-385.

[108] Mimoun, H; Chaumette, P; Mignard, M; Saussine, L; Fischer, J; Weiss, R. *Nouveau Journal de Chimie*, 1983, 7(8-9), 467-475.

[109] Tinant, B; Bayot, D; Devillers, M. *Z. Kristallogr.- New Cryst. Struct*, 2003, 218(4), 477-478.

[110] Hon, PK; Belford, RL; Pfluger, CE. *J. Chem. Phys.*, 1965, 43(4), 1323-1333.

[111] Amin, SS; Cryer, K; Zhang, B; Dutta, SK; Eaton, SS; Anderson, OP; Miller, SM; Reul, BA; Brichard, SM; Crans, DC. *Inorg. Chem.*, 2000, 39(3), 406-416.

[112] Crans, DC; Keramidas, AD; Mahroof-Tahir, M; Anderson, OP; Miller, MM. *Inorg. Chem.*, 1996, 35(12), 3599-3606.

[113] Scheidt, WR; Countryman, R; Hoard, JL., *J. Am. Chem. Soc.*, 1971, 93(16), 3878-3882.

[114] Melchior, M; Thompson, KH; Jong, JM; Rettig, SJ; Shuter, E; Yuen, VG; Zhou, Y; McNeill, JH; Orvig, C. *Inorg. Chem.*, 1999, 38(10), 2288-2293.

[115] Renolds, JG; Sendlinger, SC; Murray, AM; Hoffman, JC; Christou, G., *Inorg. Chem.*, 1995, 34, 5745-5752.

[116] Dewey, TM; Du Bois, J; Raymond, KN. *Inorg. Chem.*, 1993, 32(9), 1729-1738.

[117] Taylor, R; Kennard, O; Versichel, W., *Acta Crystallogr*, 1984, B40, 280.

[118] Kojima, A; Okazaki, K; Ooi, S; Saito, K. *Inorg. Chem.*, 1983, 22(8), 1168-1174.

[119] Payne, VCR; Headley, OSC; Stibrany, RT; Maragh, PT; Dasgupta, TP; Newton, AM; Holder, AA. *J. Chem. Crystallogr*, 2007, 37(4), 309-314.

[120] Fuerst, W; Gouzerh, P; Jeannin, Y. *J. Coord. Chem.*, 1979, 8(4), 237-243.

[121] Johnson, JE; Jacobson, RA. *Acta Crystallogr., Sect. B*, 1973, 29(8), 1669-1674.

[122] Janiak, C. *Dalton Trans.*, 2000, (21), 3885-3896.

[123] González-Baró, AC; Pis-Diez, R; Piro, OE; Parajón-Costa, BS. *Polyhedron*, 2008, 27(2), 502-512.

[124] Devereux, M; McCann, M; Leon, V; McKee, V; Ball, RJ., *Polyhedron*, 2002, 21(11), 1063-1071.

[125] Park, H; Lough, AJ; Kim, JC; Jeong, MH; Kang, YS. *Inorg. Chim. Acta*, 2007, 360(8), 2819-2823.

[126] Aghabozorg, H; Mohamad Panah, F; Sadr-Khanlou, E. *Analyt. Sciences: X-Ray Structure Analysis Online*, 2007, 23(7), x139-x140.

[127] Aghabozorg, H; Nemati, A; Derikvand, Z; Ghadermazi, M. *Acta Crystallogr., Section E: Structure Reports Online*, 2007, E63(12), m2921, m2921/1-m2921/15.

[128] Aghabozorg, H; Sadrkhanlou, E; Soleimannejad, J; Adams, H. *Acta Crystallogr., Section E: Structure Reports Online*, 2007, E63(6), m1760.

[129] Cousson, A; Nectoux, F; Robert, F; Rizkalla, EN. *Acta Crystallogr., Section C: Crystal Structure Communications*, 1995, C51(5), 838-840.

[130] Rafizadeh, M; Mehrabi, B; Amani, V. *Acta Crystallogr., Section E: Structure Reports Online*, 2006, E62(6), m1332-m1334.

[131] Sanyal, GS; Ganguly, R; Nath, PK; Butcher, RJ. *J. Indian Chem. Soc.*, 2002, 79(6), 489-491.

[132] Zhao, QH; Zhang, MS; Fang, RB. *Acta Crystallogr., Section E: Structure Reports Online*, 2005, E61(12), m2575-m2577.

[133] Uçar, I; Bulut, A; Karadag, A; Kazak, C. *J. Mol. Struct*, 2007, 837(1-3), 38-42.

[134] Braga, D; Bazzi, C; Maini, L; Grepioni, F. *CrystEngComm*, 1999, No Given, Article 5.

[135] Du, M; Cai, H; Zhao, XJ. *Inorg. Chim. Acta*, 2006, 359(2), 673-679.

[136] Liu, FC; Ouyang, J. *Acta Crystallogr., Section E: Structure Reports Online*, 2007, E63(10), m2557, Sm2557/1-Sm2557/7.

[137] Su, H; Wen, YH; Feng, Y. L. *Zeitschrift fuer Kristallographie - New Crystal Structures*, 2005, 220(4), 560-562.

[138] Sun, Q; Gao, Q; Zhang, W; Song, Y; Xu, Z; Su, B; Zhao, J. *J. Coord. Chem*, 2008, 61(5), 669-676.

[139] Wang, L; Duan, L; Wang, E; Xiao, D; Li, Y; Lan, Y; Xu, L; Hu, C. *Transition Met. Chem.*, 2004, 29(2), 212-215.

[140] Bedetti, R; Biader Ceipidor, U; Carunchio, V; Tomassetti, M. *Annali di Chimica*, 1976, 66(11-12), 741-752.

[141] Moody, L; Balof, S; Smith, S; Rambaran, VH; VanDerveer, D; Holder, A. A. *Acta Crystallogr., Sect. E: Struct. Rep. Online*, 2008, E64(1), m262-m263, m262/1-m262/14.

[142] Alcock, NW; Clarkson, GJ; Lawrance, GA; Moore, P. *Aust. J. Chem.*, 2004, 57(6), 565-570.

[143] Jiang, YM; Yin, ZJ., *Wuji Huaxue Xuebao*, 2001, 17, 589.

[144] Sileo, EE; Blesa, MA; Rigotti, G; Rivero, BE; Castellano, EE.

Polyhedron, 1996, 15(24), 4531-4540.

[145] 1Brandi-Blanco, MP; Choquesillo-Lazarte, D; GarcIa-Collado, CG; González-Pérez, JM; Castiñeiras, A; Niclós-Gutiérrez, J. *Inorg. Chem. Commun.*, 2005, 8(2), 231-234.

[146] Su, CC; Chiu, SY. *Polyhedron*, 1996, 15(15), 2623-2631.

[147] Bresciani-Pahor, N; Nardin, G; Bonomo, RP; Rizzarelli, E. *J. Chem. Soc., Dalton Trans.*, 1984, (12), 2625-3260.

[148] Nardin, G; Randaccio, L; Bonomo, RP; Rizzarelli, E. *J. Chem. Soc., Dalton Trans.*, 1980, (3), 369-375.

[149] Gao, HL; Yi, L; Zhao, B; Zhao, XQ; Cheng, P; Liao, DZ; Yan, SP. *Inorg. Chem.*, 2006, 45(15), 5980-5988.

[150] Prasad, TK; Rajasekharan, MV. *Polyhedron*, 2007, 26(7), 1364-1372.

[151] Prasad, TK; Rajasekharan, MV. *Inorg. Chem. Commun*, 2005, 8(12), 1116-1119.

[152] Prasad, TK; Sailaja, S; Rajasekharan, MV. *Polyhedron*, 2005, 24(12), 1487-1496.

[153] Sailaja, S; Rajasekharan, MV. *Acta Crystallogr., Sect. E: Struct. Rep. Online*, 2001, E57, (8), m341-m343.

[154] Swarnabala, G; Rajasekharan, MV. *Inorg. Chem.*, 1998, 37(7), 1483-1485.

[155] Shen, E; Lue, J; Li, Y; Wang, E; Hu, C; Xu, L. *J. Solid State Chem.*, 2004, 177(11), 4372-4376.

[156] Arnáiz, FJ; Aguado, R; Pedrosa, MR; De Cian, A; Fischer, J. *Polyhedron*, 2000, 19(20-21), 2141-2147.

[157] Sukanya, D; Prabhakaran, R; Natarajan, K. *Polyhedron*, 2006, 25(11), 2223-2228.

[158] Annibale, G; Cattalini, L; Canovese, L; Pitteri, B; Tiripicchio, A; Tiripicchio, CM; Tobe, ML. *J. Chem. Soc., Dalton Trans.*, 1986, 1101-1105.

[159] Espinet, P; Lorenzo, C; Miguel, JA; Bois, C; Jeannin, Y. *Inorg. Chem.*, 1994, 33, 2052-2055.

[160] Yamamoto, JH; Yoshida, W; Jensen, CM. *Inorg. Chem.*, 1991, 30, 1353-1357.

[161] Xue, L; Che, YX; Zheng, JM. *J. Coord. Chem.*, 2007, 60(13), 1381-1386.

[162] Allen, FH; Davies, JE; Galloy, JJ; Johnson, O; Kennard, O; Macrae, C. F; Mitchell, EM; Mitchell, GF; Smith, JM; Watson, DG. *J. Chem. Inf. Comput. Sci.*, 1991, 31(2), 187-204.

[163] MacDonald, JC; Luo, TJM; Palmore, GTR. *Cryst. Growth Des.*, 2004,

4(6), 1203-1209.

[164] Sileo, EE; Blesa, MA; Rigotti, G; Rivero, BE; Castellano, EE. *Polyhedron*, 1996, 15(24), 4531-4540.

[165] Moghimi, A; Sheshmani, S; Shokrollahi, A; Aghabozorg, H; Shamsipur, M; Kickelbick, G; Aragoni, M. C; Lippolis, V. *Zeit. fuer Anorg. und Allgem. Chemie*, 2004, 630(4), 617-624.

[166] Moghimi, A; Ranjbar, M; Aghabozorg, H; Jalali, F; Shamsipur, M; Chadah, RK. *J. Chem. Res., Synop.*, 2002, (10), 477-479, 1047-1065.

[167] Moghimi, A; Ranjbar, M; Aghabozorg, H; Jalali, F; Shamsipur, M; Chadha, KK. *Can. J. Chem.*, 2002, 80(12), 1687-1696.

[168] Ranjbar, M; Aghabozorg, H; Moghimi, A., *Acta Crystallogr*, 2002, E58, m304.

[169] Ma, C; Li, J; Zhang, R; Wang, D. *J. Organomet. Chem.*, 2006, 691(8), 1713-1721.

[170] Harrowfield, JM; Shahverdizadeh, GH; Soudi, AA. *Supramolecular Chemistry*, 2003, 15(5), 367-373.

[171] Swarnabala, G; Rajasekharan, MV. *Inorg. Chem.*, 1998, 37(7), 1483-1485.

[172] Prasad, TK; Rajasekharan, MV. *Cryst. Growth Des.*, 2006, 6(2), 488-491.

[173] D'Aléo, A; Toupet, L; Rigaut, S; Andraud, C; Maury, O. *Optical Mat.* 2008, 30(11), 1682-1688.

[174] Fernandes, A; Jaud, J; Dexpert-Ghys, J; Brouca-Cabarrecq, C. *Polyhedron*, 2001, 20(18), 2385-2391.

[175] Brayshaw, PA; Buenzli, JCG; Froidevaux, P; Harrowfield, JM; Kim, Y; Sobolev, AN. *Inorg. Chem.*, 1995, 34(8), 2068-2076.

[176] Harrowfield, JM; Lugan, N; Shahverdizadeh, GH; Soudi, AA; Thuery, P. *Eur. J. Inorg. Chem.*, 2006, (2), 389-396.

[177] Masci, B; Thuery, P. *Polyhedron*, 2005, 24(2), 229-237.

In: Chemical Crystallography   ISBN: 978-1-60876-281-1
Editors: Bryan L. Connelly, pp. 69-101   © 2010 Nova Science Publishers, Inc.

Chapter 2

SYNTHESIS AND STRUCTURAL CHARACTERIZATION OF THIOPHENE-FUNCTIONALIZED METAL DITHIOLENES

Seth C. Rasmussen and *Chad M. Amb*

Department of Chemistry and Molecular Biology, North Dakota State
University, NDSU Dept. 2735, P.O Box 6050, Fargo, ND 58108, USA

INTRODUCTION

Transition metal dithiolenes are versatile complexes capable of a wide range of oxidation states, coordination geometries, and magnetic moments.[1] As a consequence, these complexes have been widely studied as building blocks for crystalline molecular materials. Particularly successful are the square-planar metal dithiolenes (Chart 1), from which materials have been produced that exhibit conducting, magnetic, and nonlinear optical properties, as well as superconductivity in some cases.[1-3] In their application to molecular-based conductors, metal dithiolenes can play several different roles. Metal dithiolenes may form an effective conduction pathway through intermolecular face-to-face stacking, or can play a supportive role as counterions to other planar molecules (such as perylene or tetrathiafulvalene derivatives, Chart 1) which provide the actual conduction path.[3] When acting as counterions, such

_____

* Correspondig author: Email: seth.rasmussen@ndsu.edu.

dithiolene complexes can additionally impart magnetic properties to the molecular conductors via interactions between localized spins of the metal dithiolene with the intinerant spins of conduction electrons.

As a result of such research efforts, there now exists a variety of available dithiolene ligands which have been applied to produce a broad range of metal dithiolene materials. One focus of study has been the use of electronically delocalized dithiolene ligands to explore the influence on the solid-state structures and the resulting material properties.[2] Of particular interest has been the preparation and study of metal dithiolenes functionalized with thiophene moieties. The application of such extended π-systems and sulfur-rich ligands are expected to enhance solid-state interactions, which could result in enhanced electrical conductivity or higher magnetic transition temperatures. As the molecular packing in the crystal is determined by the total balance of many weak intermolecular forces (hydrogen bonding, van der Waals, π–π interactions, and S⋯S/M⋯S interactions),[2,3] the additional thiophene content would increase such intermolecular interactions and provide more significant overlap of frontier orbitals. In addition, such complexes could provide potential precursors to metal-dithiolene-containing conjugated polymers.

Herein, we will review the preparation and study of this class of functionalized metal dithiolenes, with particular emphasis on the relationships between functionality and solid-state structure, and the overall effect on material properties. Discussion will go beyond the application of this class of thiolene complexes as molecular materials, and equal attention will be given to the effectiveness of various structural motifs when designing these species as polymeric precursors.

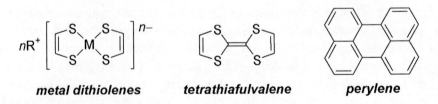

metal dithiolenes **tetrathiafulvalene** **perylene**

Chart 1. Metal dithiolenes and other common building blocks of molecular-based conductors.

Metal 2,3-Thiophenedithiolene Complexes

Synthesis. The first known thiophene-based metal dithiolenes were reported by Gol'dfarb and Kalik in 1968.[4] These consisted of the nickel and copper complexes of 2,3-thiophenedithiolate produced from ethyl 3-thienyl sulfide as shown in Scheme 1. Both complexes were reported as black, high melting solids and were thought to be the neutral bis(2,3-thiophenedithiolate) species.

Such 2,3-thiophenedithiolene complexes were then prepared through an alternate route by Henriques, Almeida and coworkers in 2001 (Scheme 2).[5-7] Here, the dithiolate was generated from the precursor thieno[2,3-*d*]-1,3-dithiol-2-one[7] and then reacted with the corresponding metal salts without isolation of the reactive intermediate. While the nickel complex was also prepared here, and the reaction with the NiCl$_2$ salt was again carried out under air, elemental analysis confirmed that oxidation did not proceed to the neutral species, but generated the monoanion **2a** via a single-electron oxidation.[6] Recrystallization of [Bu$_4$N][**2a**] from acetone/2-propanol gave dark-green, plate-like crystals which decomposed at 130 °C,[3] well below the 360 °C melting point reported for **1a** by Gol'dfarb and Kalik.[2]

While initial salts of the Ni and Au thiophenedithiolene complexes were made as tetrabutylammonium (Bu$_4$N$^+$) salts, difficulties in crystallizing the resulting salts led to the use of larger cations such as tetraphenylphosphonium (Ph$_4$P$^+$) for later complexes of Co and Pt.[8] A further modification was that the preparation of these complexes was carried out in strictly anaerobic conditions through the use of a glovebox. As a result, the dianionic complexes [Ph$_4$P]$_2$[**3a**] and [Ph$_4$P]$_2$[**3b**] were isolated as dark brown and orange solids respectively.[8] In an attempt to generate the monoanionic form of **3b**, the dianionic complex was dissolved in acetonitrile and exposed to air. This resulted in a color change from orange to green, consistent with oxidation of the **3b** anion, but isolation of the monoanionic species was not successful.

Scheme 1. Original synthesis of metal 2,3-thiophenedithiolate complexes.

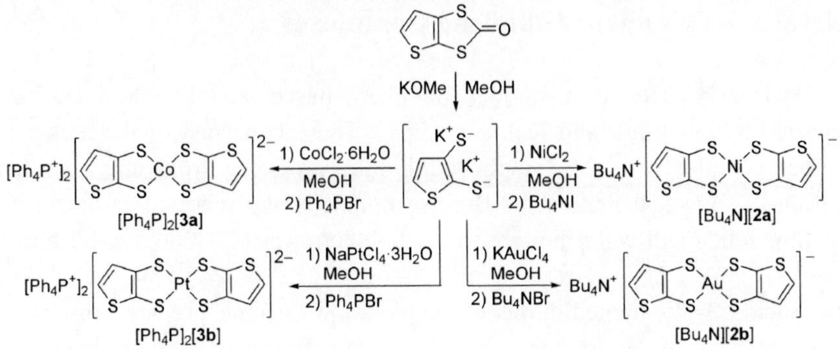

Scheme 2. Modern synthesis of metal 2,3-thiophenedithiolenes.

In addition to the cations discussed above, additional salts of the monoanion **2a** have been prepared by substituting various pyridinium salts for the Bu$_4$NI utilized in Scheme 2.[9] A significant number of salts of **2a** have also been prepared via metathesis of [Bu$_4$N][**2a**] with various metal PF$_6$ salts. These methods have thus produced salts of **2a** with various organometallic cations ([M(Cp*)$_2$]$^+$, were M = Fe, Mn, Co, Cr).[6,10]

X-ray Crystallography. All of the salts described above can be crystallized to produce well-formed single crystals and the majority have been studied by X-ray crystallography (Table 1). While well-formed crystals of the parent [Bu$_4$N][**2a**] are easily obtained, its crystal structure could not be solved due to severe twinning problems. The use of bulkier cations, however, has been more successful and the ellipsoid plot of the complex **2a** from the salt [Fe(Cp*)$_2$][**2a**] is shown in Figure 1. Selected bond distances of **2a** from the various reported structures are given in Table 2.[6,9,10] The Ni-S bond lengths observed in the various structures of **2a** range from 2.154 to 2.173 Å, consistent with other known Ni dithiolene complexes.[1] The exterior fused thiophene rings agree fairly well with the structure of the parent thiophene,[11] with the exception of a small amount of asymmetry as previously reported for other thiophenes fused along the *b*-face of the heterocycle.[12] The length of the **2a** anion is nearly planar with only a minor tetrahedral distortion of the Ni coordination geometry due to a small twist (~3-6°) of one dithiolate relative to the other along the long axis of the complex. Such small distortions are typical of square planar metal dithiolenes.[1]

Table 1. Crystal cell parameters for various salts of 2a[a].

Cation	[Bu$_4$N]$^+$	[Fe(Cp*)$_2$]$^+$	[Co(Cp*)$_2$]$^+$	[Cr(Cp*)$_2$]$^+$	[BzPy]$^+$	[BrBzPy]$^+$	[FBzPy]$^+$
Crystal System	monoclinic	monoclinic	monoclinic	monoclinic	monoclinic	monoclinic	orthorhombic
Space Group	NR	P2$_1$/a	NR	P2$_1$/c	P2$_1$/c	P2$_1$/c	P2$_1$2$_1$2$_1$
a (Å)	9.4801(19)	15.443(3)	15.370(3)	10.046(3)	14.9140(9)	15.8727(2)	10.4280(12)
b (Å)	18.7867(24)	10.237(1)	10.233(2)	10.270(3)	9.3980(7)	8.30210(10)	12.3940(6)
c (Å)	9.6185(20)	20.360(2)	20.450(3)	15.528(3)	15.3170(10)	16.7073(2)	16.5940(19)
α (°)	90.00	90.00	90.00	90.00	90.00	90.00	90.00
β (°)	119.336(23)	107.54(1)	108.41(3)	104.88(2)	97.784(4)	90.2790(10)	90.00
γ (°)	90.00	90.00	90.00	90.00	90.00	90.00	90.00
V (Å^3)	1493(1)	3069.1(7)	3051.8	1548.3(7)	2127.1(15)	2201.61(5)	2144.7(4)
Z	NR	4	NR	2	4	4	4
CCDC deposit no.	b	152949	c	615973	600733	600731	600732
Reference	6	5,6	6	10	9	9	9

[a]NR = not reported. [b]not solved due to twinning. [c]not refined due to low quality crystals.

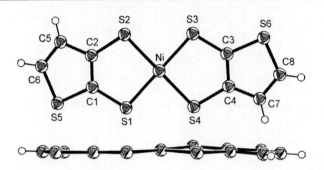

Figure 1. Face and edge ellipsoid plot of the complex **2a** from the salt [Fe(Cp*)₂][**2a**] at the 50% probability level.

Table 2. Selected bond distances (Å) of 2a with various cations.

Parameter	[Fe(Cp*)₂][2a][a]	[Cr(Cp*)₂][2a][b]	[BzPy][2a][c]	[BrBzPy][2a][c]	[FBzPy][2a][c]
Ni–S1	2.154(4)	2.169(2)	2.1628(9)	2.1732(7)	2.173(1)
Ni–S2	2.162(3)	2.159(2)	2.1633(8)	2.1673(7)	2.162(1)
S1–C1	1.70(1)	1.716(6)	1.717(3)	1.723(3)	1.725(5)
S2–C2	1.74(1)	1.737(7)	1.733(3)	1.723(3)	1.740(5)
C1–C2	1.39(2)	1.362(9)	1.377(4)	1.377(4)	1.367(7)
C1–S5	1.72(1)	1.732(7)	1.733(3)	1.707(5)	1.731(5)
C2–C5	1.46(2)	1.474(9)	1.491(4)	1.47(2)	1.460(7)
C5–C6	1.36(2)	1.43(1)	1.374(5)	1.37(2)	1.362(8)
S5–C6	1.70(1)	1.697(7)	1.705(3)	1.559(7)	1.706(6)
[a]Reference 6. [b]Refence 10. [c]Reference 9.					

In the case of the salts of organometallic cations, the **2a** anion adopts only a *trans* configuration in the crystal structures.[6,10] However, in all of the pyridinium salts, there is clear evidence for variable amounts of *cis–trans* disorder,[9] which is somewhat unusual for such thiophenedithiolene complexes. The trans configuration is still predominate for structures of the [BzPy]⁺ and [BrBzPy]⁺ salts, while the structure of the [FBzPy]⁺ salt exhibits ~77% *cis* conformation. In addition to the *cis–trans* disorder, the **2a** anion in the [BrBzPy]⁺ salt also exhibits ~20% orientation disorder.[9]

The crystal structure of [Fe(Cp*)₂][**2a**] consists of alternating layers of [Fe(Cp*)₂]⁺ cations and **2a** anions (Figure 2A), with each layer lying along the *a,b* plane.[6] The anion layers contain short intermolecular C–H···S contacts of 2.964 Å, slightly less than the sum of the van der Walls radii,[13] with corresponding C–H···S angles of 172.18° (Figure 2B). Such interactions are characteristic of weak hydrogen bonding[14] and provide ribbons of hydrogen-bonded **2a** complexes throughout the anion layers. Cofacial close contacts (~3.50 Å) are also seen between the aromatic faces of the thiophene moieties of **2a** and the Cp* ligands of adjacent cation layers. Such interactions are consistent with strong π-stacking[15] and result in the formation of extended π-stacked columns of alternating anions and cations, as shown in Figure 3.

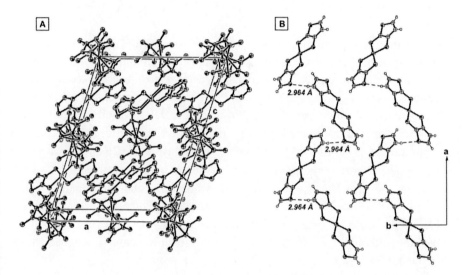

Figure 2. Crystal packing of [Fe(Cp*)$_2$][**2a**] (A) and close-contacts between **2a** anions (B).

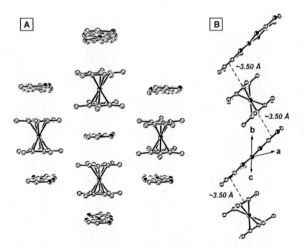

Figure 3. Cofacial π-stacked columns of alternating [Fe(Cp*)$_2$]$^+$ cations and **2a** anions.

While $[Co(Cp^*)_2][2a]$ is thought to be isostructural with $[Fe(Cp^*)_2][2a]$,[6] $[Cr(Cp^*)_2][2a]$ adopts a slightly different, but very similar supramolecular packing.[10] Unlike $[Fe(Cp^*)_2][2a]$, no contacts shorter than the sum of the van der Waals radii were detected in the structure of $[Cr(Cp^*)_2][2a]$. However, π-stacking between the thiophene rings of the 2a anions and the cation Cp* ligands were again observed, with a slightly larger ring to ring separation of 3.576 Å.[10]

In the three pyridinium salts, the crystal structures lack the extended π-stacking of the previous structures and seem to be dominated by the packing pattern of the 2a anions.[9] The crystal structures are generally composed of double layers of anions. These double layers are then separated by layers of the cations, in which one aromatic ring of the cation inserts itself into the anion layer via π-interactions and the second aromatic ring forms the separating layer (Figure 4A).[9] Within the anion layers, the 2a anions tend to be positioned almost perpendicular to each other, with intermolecular C–H$\cdots$S contacts between the α-position of the thiophene and coordinating sulfur atoms as illustrated in Figure 4B for [BzPy][2a]. Some close S$\cdots$S contacts (3.497 Å) are also observed. The remaining two structures exhibit similar packing, although with some differences in the close contacts between anions in the double layers.[9]

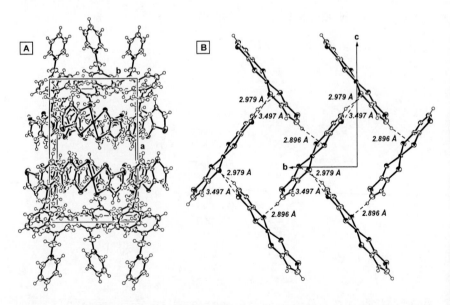

Figure 4. Crystal packing of [BzPy][2a] (A) and close-contacts between 2a anions (B).

In addition to the crystal structures of the various salts of the monoanion **2a** discussed above, the structures of the Co and Pt dianions [Ph$_4$P]$_2$[**3a**] and [Ph$_4$P]$_2$[**3b**] have also been reported.[8] These dianions exhibited average Co–S and Pt–S bond lengths of 2.218 and 2.310 Å, respectively, and are consistent with other known Co and Pt dithiolene complexes.[1] The bond lengths of the 2,3-thiophenedithiolate ligands all agree well with the previous structures of **2a**. As with the pyridinium salts discussed above, both the **3a** and **3b** dianions exhibit *cis-trans* disorder, making it impossible to distinguish between a *cis* or *trans* conformation of the two complexes.[8]

Solid State Properties. A number of the salts of the S = 1/2 **2a** anion exhibit magnetic ordering. Characterization of [Fe(Cp*)$_2$][**2a**] at low applied magnetic fields revealed an antiferromagnetic transition at T_N = 2.56 K. Below 2.6 K, [Fe(Cp*)$_2$][**2a**] displays a metamagnetic behavior with a critical field H_C of 70 mT at 1.6 K. Above 7 T, the magnetization becomes almost saturated, attaining a value of 2.45 μ_B/mol at 12 T.[6] Such metamagnetism can often arise in low-dimensional systems containing strong ferromagnetic interactions in one dimension and weak antiferromagnetic interactions between chains. The applied field can then overcome the weak antiferromagnetic interactions and align the chains. Here the observed metamagnetic behavior is believed to be largely due to the cation-anion π-stacking interactions seen in the crystal structure. This is supported by the study of the isostructural [Co(Cp*)$_2$][**2a**], which contains a diamagnetic cation and thus removes any anion-cation magnetic interactions. Characterization of [Co(Cp*)$_2$][**2a**] revealed only antiferromagntic behavior, thought to be the result of anion-anion interactions.[6]

The additional analogues [Cr(Cp*)$_2$][**2a**] and [Mn(Cp*)$_2$][**2a**] were later studied and the McConnell I model was used to analyze the intermolecular magnetic coupling in this family of [M(Cp*)$_2$][**2a**] salts.[10] The predicted ferromagnetic anion-cation coupling and the antiferromagnetic coupling in the anion-anion coupling (both intra and interlayer) were in good agreement with the experimental results. The magnetic behavior of these compounds is thought to be dominated by ferromagnetic interactions, which are ascribed to the anion-cation interactions. The low-temperature magnetic behavior of [Cr(Cp*)$_2$][**2a**] and [Mn(Cp*)$_2$][**2a**] contrasts with that previously reported for [Fe(Cp*)$_2$][**2a**]. Whereas [Cr(Cp*)$_2$][**2a**] remains paramagnetic down to 1.6 K, [Mn(Cp*)$_2$][**2a**] shows magnetic behavior that is typical of a frustrated magnet and has a blocking temperature of ~4 K. For this compound, the magnetic frustration is reported to result from a degenerate ground state in the interlayer spin arrangements.[10] The differences in the magnetic properties within this family are thought to be due to differences in both the π-stacking anion-cation interactions as well as differences in the anion-anion close contacts. While the pyridinium salts lack any possible anion-cation-based magnetic interactions, these compounds still exhibit magnetic exchange as a result of close anion-anion contacts.[9] The salt BzPy[2a] displays dominant ferromagnetic interactions and cluster glass behavior at low temperatures, while BrBzPy[2a] exhibits dominant antiferromagnetic interactions with a magnetic anomaly at ~6 K. The magnetic behavior of the final analogue FBzPy[2a] is dominated by weak ferromagnetic interactions, but without magnetic ordering down to 1.5 K.[9]

In addition, the oxidation of [Bu$_4$N][2b] resulted in a neutral complex as a polycrystalline sample. This material displays metallic properties with a room temperature electrical conductivity of 6 S cm^{-1} and a thermoelectric power of 5.5 μV K^{-1}.[7] This is reported to be the first example of such metallic properties observed in a molecular system based on a neutral species.[7]

Extended Metal 2,3-Thiophenedithiolene Complexes

Synthesis. In attempts to further extend the π-system of the metal dithiolene core of 2a, Amb and Rasmussen reported the preparation of a 2,2'-bithiophenedithiolate ligand to synthesize [Bu$_4$N][4a] (Scheme 3) in 2007.[16] The resulting anion 4a represents a functionalized analogue of 2a, with the π-system extended through the α-thienyl groups coupled to the exterior α-positions of the 2a core. This initial synthetic approach was then optimized[17] and used to produce a series of extended metal thiophenedithiolene complexes 4a-e as shown in Scheme 3.[18]

The synthesis of the desired thiophenedithiolate ligands was achieved through regioselective cross-coupling of 2,3,5-tribromothiophene to generate 5-aryl-2,3-dibromothiophene intermediates.[19] These intermediates could then be converted to the corresponding thiophenedithiolates through sequential alkyl lithium and sulfur additions, followed by treatment with acetylchloride to generate the stable acetyl-protected thiophenedithiolates.[17,18] The acetyl-protected 5-(2-thienyl)-2,3-thiophenedithiolate could be further modified through regioselective electrophilic substitutions to give the bromo- and acetyl-derivatized ligands.[18]

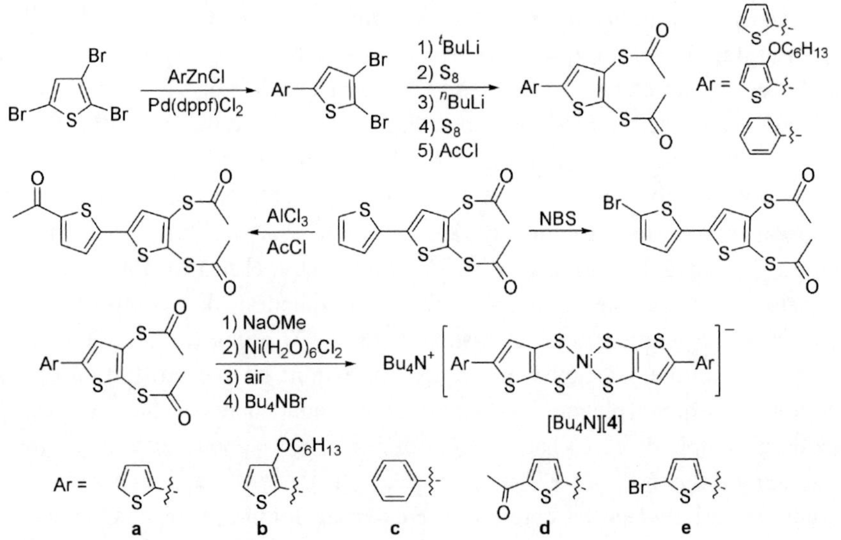

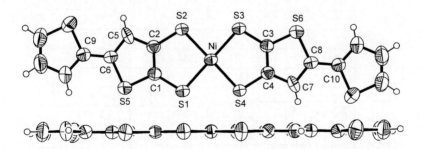

Scheme 3. Synthesis of metal 5-aryl-2,3-thiophenedithiolenes.

The corresponding nickel complexes of these ligands were synthesized through deprotection of the acetyl groups via saponification in methanol, followed by addition of $Ni(H_2O)_6Cl_2$ and air oxidation. The monoanions **4a-e** were then precipitated by treatment with tetrabutylammonium bromide, and recrystallized to give shiny, crystalline materials. N-Methylpyridinium salts were also produced by substituting N-methylpyridinium iodide for tetrabutylammonium bromide.[17,18]

Figure 5. Face and edge ellipsoid plot of **4a** from the salt [MePy][**4a**] at the 50% probability level.

X-ray Crystallography. The *N*-methylpyridinium salts [MePy][**4a**], [MePy][**4c**], and [MePy][**4e**] have been studied by X-ray crystallography and their crystal structures determined.[17,18] An ellipsoid plot of the complex **4a** from the salt [MePy][**4a**] is shown in Figure 5 and selected bond distances for the Ni thiophenedithiolene core of **4a**, **4c**, and **4e** are given in Table 3.[18] The Ni thiophenedithiolene cores of the three structures are consistent with the previously reported structure of **2a**,[6,9,10] with observed Ni–S bond lengths ranging from 2.167 to 2.176 Å. As previously observed for the various structures of **2a**, **4e** exhibits a minor tetrahedral distortion of the Ni coordination geometry due to a small twist (~7°) along the long axis of the complex. In contrast, however, **4a** and **4c** exhibit no such distortion and are completely square planar.[18] The interannular bond between the core and the external thiophene of **4a** has a length of 1.445 Å, in good agreement with the interannular bond of 2,2'-bithiophenes (1.45 Å)[20] and suggests good conjugation between the core and the external thiophene. In comparison, the interannular bond of **4c** is elongated, potentially suggesting reduced conjugation between the phenyl and thiophene rings. This could also explain some of the significant differences observed within the bond lengths of the thiophene moiety for this complex.

Table 3. Selected bond distances (Å) of [MePy][4a], [MePy][4c], and [MePy][4e][a].

Parameter	[MePy][**4a**]	[MePy][**4c**]	[MePy][**4e**]
Ni–S1	2.167(2)	2.174(3)	2.167(1)
Ni–S2	2.171(2)	2.176(3)	2.170(1)
S1–C1	1.714(6)	1.732(9)	1.717(5)
S2–C2	1.732(5)	1.731(10)	1.735(5)
C1–C2	1.382(6)	1.442(12)	1.371(6)
C1–S5	1.733(5)	1.725(8)	1.746(5)
C2–C5	1.486(9)	1.439(13)	1.446(7)
C5–C6	1.423(8)	1.395(11)	1.393(7)
S5–C6	1.745(5)	1.739(8)	1.747(5)
C6–C9	1.445(9)	1.51(1)	1.441(7)
[a]Reference 18.			

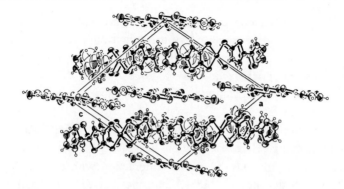

Figure 6. Crystal packing of [MePy][**4a**].

As shown in Figure 6, the crystal packing of **4a** consists of alternating layers of π-stacked anion-cation columns and 2-D anion-anion sheets with strong edge-to-edge interactions.[17,18] The sheets comprised of anion-anion contacts are detailed in Figure 7A, with the anions here exhibiting a planar Ni thiophenedithiolene with a 10° twist about the interannular bond to the external thiophenes. The edge-to-edge sheets are ordered via complimentary close contacts mediated by sulfur atoms, with two C–H···S (2.966 Å) and one S···S (3.592 Å) interactions between each pair of anions.[17,18]

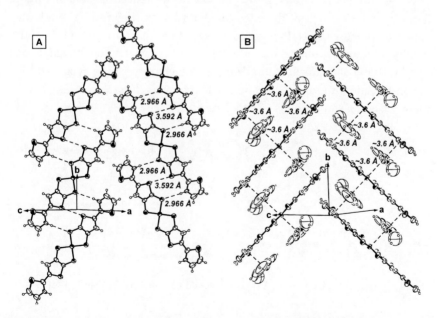

Figure 7. Edge-to-edge sheets (A) and π-stacked columns (B) of [MePy][**4a**].

The π-stacked columns are detailed in Figure 7B and are comprised of alternating sets of one **4a** anion to two methylpyridinium cations, and thus all of the cations in the crystal reside within these π-stacked columns. In contrast to the anions comprising the edge-to-edge sheets, the anions here are completely planar (<1° deviation). The face-to-face distance between anions and cations is ~3.6 Å.[17,18]

In contrast to [MePy][**4a**], its brominated analogue [MePy][**4e**] exhibits no extended π-stacking. Here each anion has one S⋯S close contact with a neighboring anion and one π-stacking interaction with a methylpyridinium cation. The S⋯S distance is 3.385 Å and the face-to-face distance between anion and cation is ~3.4 Å.[18]

More similar to [MePy][**4a**], the phenyl derivative [MePy][**4c**] again shows extended arrays as a result of S⋯S and π-stacking interactions. The primary difference here is that rather than separate layers of edge-to-edge sheets and π-stacked columns seen in **4a**, the anions of **4c** exhibit S⋯S (3.546 Å) and C–H⋯S (2.968 Å) to generate extended edge-to-edge ribbons as shown in Figure 8A. These ribbons are not isolated to layers as in the structure of [MePy][**4a**] and multiple ribbons all run perpendicular to one another. In addition, face-to-face π-stacking between the anions of these ribbons and the methylpyridinium cations result in extended π-stacked columns running perpendicular to each of the edge-to-edge ribbons (Figure 8B). While the π-stacked columns in [MePy][**4a**] consisted of alternating layers of **4a** and two methylpyridinium cations, the columns seen here contain only a single cation per layer and an equivalent of DMF fills the empty position generated by this anion-to-cation ratio in the π-stacked columns. The face-to-face distance between anion and cation is ~3.5-3.7 Å.[18]

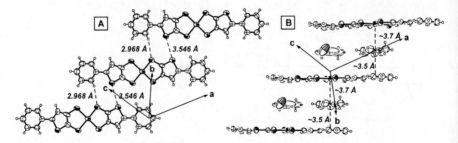

Figure 8. Edge-to-edge sheets (A) and π-stacked columns (B) of [MePy][**4c**].

Solid State Properties. The conductivity of the MePy$^+$ and Bu$_4$N$^+$ salts of complexes **4a-e** have been measured by the four-point probe technique and with the exception of [MePy][**4a**], all other salts were found to be insulating. The complex [MePy][**4a**] exhibited a pressed pellet conductivity of 8.8 x 10^{-6} S/cm, indicating that the compound is a molecular semiconductor.[17,18] This result is interesting as dithiolene complexes in integral charge states (-2 or -1) generally show insulating behavior and show paramagnetism or magnetic coopertivity, while dithiolenes complexes in non-integral charge states often give conductive materials.[2] Another salt, [Mn(tpy)$_2$][**4a**]$_2$ (where tpy = 2, 2':6',2"-terpyridine) exhibited a conductivity of 7.7 x 10^{-7} S/cm.[18] A solution of [MePy][**4a**] in CH$_2$Cl$_2$ was also oxidized by treatment with I$_2$, but the conductivity of pressed pellets was actually reduced from that of the salt, suggesting a highly disordered material.

The magnetic properties of [MePy][**4a**] were also investigated to give somewhat unexpected results. As a −1 nickel dithiolene with similar solid-state packing and interactions to the previous salts of **2a**, it would be expected to exhibit paramagnetism. However, while solution measurements show that the complex **4a** is paramagnetic with μ = 1.9 μB, [MePy][**4a**] was shown to be diamagnetic in the solid state.[17] It was hypothesized that the excellent electronic communication throughout the bulk solid which allowed conductivity through the material also resulted in antiferromagnetic coupling of spins, which resulted in the diamagnetism at room temperature.

Metal 3,4-Thiophenedithiolene Complexes

Synthesis. In addition to the 2,3-thiophenedithiolene complexes discussed above, Almeida and coworkers also investigated the isomeric 3,4-thiophenedithiolene complexes, beginning with the Au complex in 2001 (Scheme 4, [Bu$_4$N][**5a**]).[5,7] Analogous to the synthesis of the metal 2,3-thiophenedithiolenes, the dithiolate ligand was generated from thieno[3,4-*d*]-1,3-dithiol-2-one[21] and reacted with the corresponding metal salts without isolation of the reactive intermediate. As before, a move to the larger tetraphenylphosphonium (Ph$_4$P$^+$) cation and the application of strictly anaerobic conditions resulted in the isolation of the dianionic Pt and Co complexes, [Ph$_4$P]$_2$[**6a**] and [Ph$_4$P]$_2$[**6b**], as red and green solids respectively.[8] The dianionic **6b** was dissolved in acetonitrile and stirred under air to give the monoanionic [Ph$_4$P][**5b**] as a dark blue precipitate.

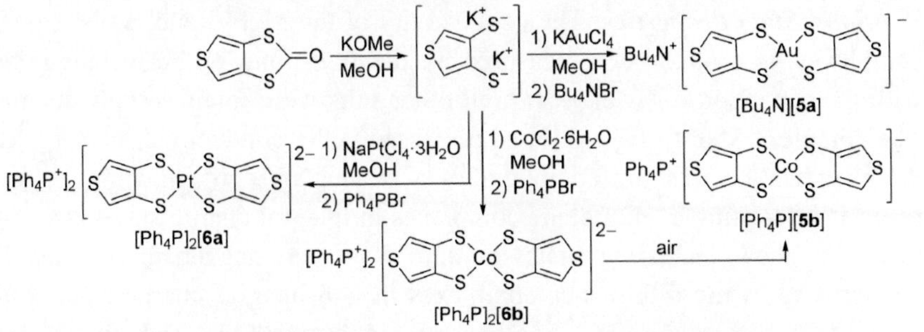

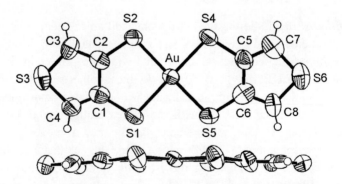

Scheme 4. Synthesis of metal 3,4-thiophenedithiolenes.

X-ray Crystallography. The metal 3,4-thiophenedithiolenes [Bu$_4$N][**5a**], [Bu$_4$N][**5b**] and [Bu$_4$N]$_2$[**6a**] have been studied by X-ray crystallography and their crystal structures determined.[7,8] An ellipsoid plot of the complex **5a** is shown in Figure 9 and selected bond distances of the metal thiophene-dithiolene core of **5a**, **5b**, and **6a** are given in Table 4.[7,8] The anions **5b** and **6a** exhibit inversion centers at the metal centers and the M–S bond lengths of all three anions agree with those seen in previous dithiolenes of the respective metals.[1] As with the previous 2,3-thiophenedithiolene complexes, the exterior fused thiophene rings of the 3,4-thiophenedithiolene complexes agree fairly well with the structure of the parent thiophene.[11] Unlike many of the previous structures, none of the anions exhibited any tetrahedral distortion of the metal coordination geometry and all three structures are almost planar, with only small deviations observed at the dithiolene sulfurs which cause the thiophenes to bend slightly out of plane.[7,8]

Figure 9. Face and edge ellipsoid plot of **5a** from the salt [Bu$_4$N][**5a**] at the 50% probability level.

Table 4. Selected bond distances (Å) of [Bu₄N][5a],
[Bu₄N][5b] and [Bu₄N]₂[6a].

Parameter	[Bu₄N][5a][a]	[Ph₄P][5b][b]	[Ph₄P]₂[6a][c]
M–S1	2.315(5)	2.177(1)	2.3109(7)
M–S2	2.302(5)	2.182(1)	2.3114(6)
S1–C1	1.75(3)	1.749(4)	1.759(2)
S2–C2	1.78(2)	1.760(4)	1.754(3)
C1–C2	1.42(3)	1.421(6)	1.436(3)
C1–C4	1.34(3)	1.360(5)	1.363(4)
C2–C3	1.35(3)	1.334(6)	1.370(3)
S3–C3	1.72(3)	1.710(4)	1.721(2)
S3–C4	1.70(2)	1.709(4)	1.714(3)
[a]M = Au, Reference 7. [b]M = Co, Reference 8. [c]M = Pt, Reference 8.			

The structure of [Bu₄N][**5a**] consists of alternating layers of Bu₄N⁺ cations and edge-to-edge sheets of **5a** anions parallel to the *ab* plane (Figure 10A). The edge-to-edge sheets comprised of anion-anion contacts are detailed in Figure 10B. The anions comprising these sheets are ordered via complimentary close contacts mediated by sulfur atoms, with one C–H···S (2.931 Å) and S···S (3.470 Å) interaction between each pair of anions.[7]

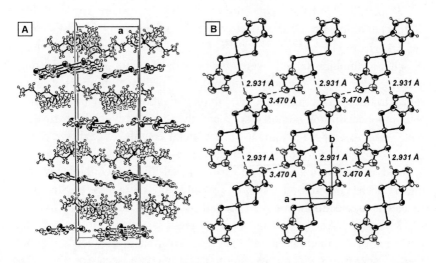

Figure 10. Crystal packing of [Bu₄N][**5a**] (A) and close-contacts between **5a** anions (B).

The crystal structure of [Ph$_4$P][**5b**] is similar, although with fewer edge-to-edge contacts within the anion layers. Chains of **5b** anions are still connected between the terminal sulfur atoms by short S···S (3.492 Å) interactions. However, rather then extended edge-to-edge interactions to form sheets, these chains are interconnected by additional **5b** anions placed almost perpendicularly to the chains. These bridging anions share short S···C (3.436(2) Å) contacts with the chains, which may denote a weak interaction between the S atom of the bridging anion and the π system of the extended chains. In contrast to the other two structures, the structure of [Bu$_4$N]$_2$[**6a**] consists of isolated anions and exhibits no anion-anion contacts.[8]

Solid State Properties. Although complex [Bu$_4$N][**5b**] is diamagnetic, it was successfully oxidized with iodine to the paramagnetic compound [Bu$_4$N][**5b**]$_{-2}$, which was isolated as a fine powder. Static magnetic susceptibility measurements of this compound over the temperature range 4-300 K confirm its paramagnetic behavior. However, the paramagnetic susceptibility, calculated from the raw measurements after correction for the diamagnetism estimated from tabulated Pascal constants, is rather small (5 × 10^{-4} emu mol^{-1}) at room temperature.[7]

Extended Metal 3,4-Thiophenedithiolene Complexes

Synthesis. Further extension of the π-system of metal 3,4-thiophenedithiolenes was successfully accomplished by Skabara and coworkers beginning in 2002.[22] Here the α-positions of the 3,4-thiophenedithiolate ligand were functionalized with 2-thienyl groups to generate a new 3',4'-terthiophenedithiolate ligand. As with previous synthetic routes to thiophenedithiolenes, the dithiolate ligand was generated *in situ* from a terthiophenedithiolone precursor[23] and then reacted with the corresponding metal salts to prepare a number of new metal complexes (Scheme 5).[22-24] As these complexes were prepared under argon via Schlenk techniques, **7a** was isolated as the dianion similar to various metal thiophenedithiolenes prepared under drybox conditions above.[24] While air oxidation of **7a** was not investigated, electrochemistry of the salt [Bu$_4$N]$_2$[**7a**] exhibited an oxidation at -0.13 V vs. Ag/AgCl.[22,24] This redox couple was assigned to the one electron oxidation of the dianion **7a** and thus the corresponding monoanion should be readily accessible via air oxidation as previously seen for the other Ni thiophenedithiolenes.

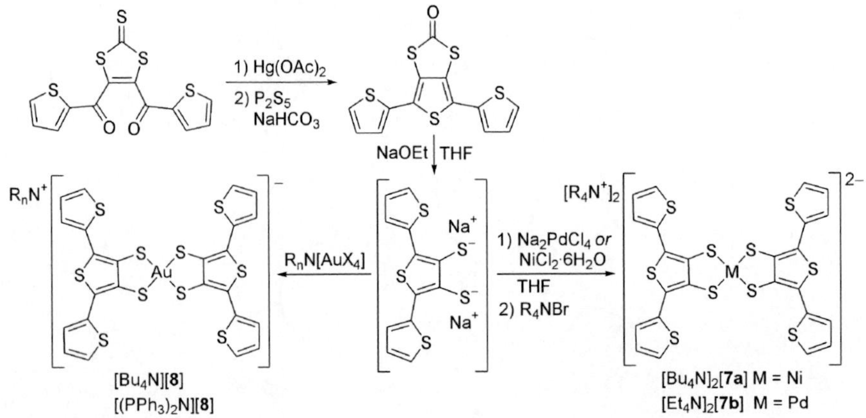

Scheme 5. Synthesis of metal 2,5-bis(2-thienyl)-3,4-thiophenedithiolenes.

Using analogous synthetic methods, the methyl-capped derivative of [Bu$_4$N]$_2$[**7a**] and a series of metal 2,5-bis(*p*-methoxyphenyl)-3,4-thiophene-dithiolenes were also prepared (Chart 2). While the Ni complex [Et$_4$N][**10a**] was prepared under the same conditions as [Bu$_4$N]$_2$[**7a**], it was surprisingly isolated as the monoanion, rather than the dianion as previously discussed. However, unlike the previous methods, the purification of [Et$_4$N][**10a**] involved water washes to remove excess [NEt$_4$]Br.[24] Such water treatment could easily explain the oxidation of an initially formed dianion to the isolated monoanionic species.

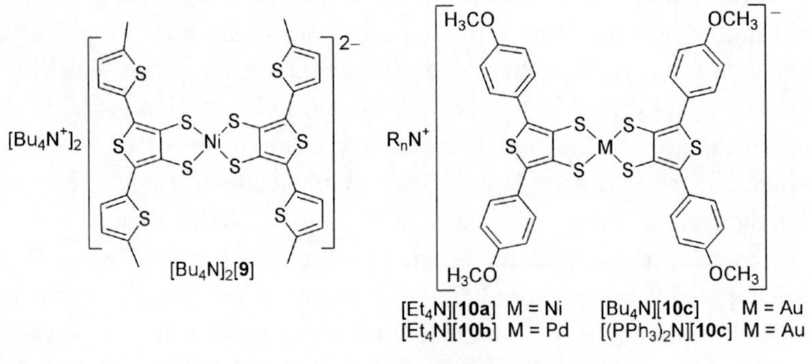

Chart 2. Additional metal 2,5-bisaryl-3,4-thiophenedithiolenes.

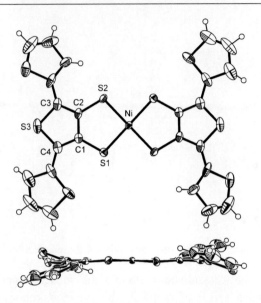

Figure 11. Face and edge ellipsoid plot of the complex **7a** from the salt [Bu₄N]₂[**7a**] at the 50% probability level.

X-ray Crystallography. The complexes [Bu₄N]₂[**7a**], [Bu₄N]₂[**9**], [(PPh₃)₂N][**8**] and [(PPh₃)₂N][**10c**] have all been successfully crystallized and their structures determined via X-ray crystallography.[22,24] An ellipsoid plot of the complex **7a** is shown in Figure 11 and selected bond lengths of the Ni thiophenedithiolene core of **7a** and **9** are given in Table 5.[22,24] The anions **7a** and **9** exhibit inversion centers at the nickel metal centers, with observed Ni–S bond lengths ranging from 2.181 to 2.195 Å. While these bonds are a bit longer than seen for the other Ni thiophenedithiolenes above, the M-S bond lengths of metal dithiolenes roughly correlate in an inverse fashion with the complex charge[1] and thus the dianions here would be expected to give longer bond lengths than the previously discussed monoanions. These values are still within the common range of Ni–S bond lengths for Ni dithiolenes.[1]

In contrast to many of the previous complexes, neither **7a** nor **9** exhibit any distortion of the Ni coordination geometry about the Ni center and are completely square planar.[18] The exterior thiophene rings, however, crystallize out of plane in relation to the thiophenedithiolene core, thus removing planarity from the anion as a whole. In addition, positional disorder in the peripheral thiophene rings leads to distributions of all-*anti* and *anti–syn* conformations within the terthiophene units. The dihedral angles observed between the exterior thiophenes and the thiophenedithiolene core range between 4.5 and

26.2°, with corresponding interannular bond lengths ranging from 1.441 to 1.452 Å. Even with the nonplanarity of the terthiophene units, these bond lengths are still in good agreement with the interannular bond of 2,2'-bithiophenes (1.45 Å),[20] thus suggesting significant conjugation along the terthiophene backbone.

Short intramolecular S···S contacts ranging between 3.159 and 3.223 Å are observed between the external thiophenes and the coordinating sulfur atoms of the dithiolene. These contacts are shorter than the sum of the corresponding van der Waals radii (3.60 Å)[13] and shorter than the various S···S contacts exhibited by the previous structures discussed above. However, unlike all of the previous structures, no significant intermolecular interactions were observed between the metal thiophenedithiolene anions of **7a** or **9**.[22,24]

The two gold thiophenedithiolenes give crystal structures with very similar structural features. The average Au–S bond lengths exhibited are 2.327 Å for **8** and 2.303 Å for **10c**,[24] consistent with previous Au dithiolenes.[1] As with the Ni complexes, no significant intermolecular short contacts are observed.[24]

Solid State Properties. Perhaps as a result of the limited intermolecular interactions of these complexes in the solid state, they exhibit few solid-state properties of interest. The dianionic **7a** did successfully produce a charge transfer material by mixing it with a solution of the tetrathiafulvalene-based salt $(TTF)_3(BF_4)_2$. Elemental analysis of the resulting black product gave the formula $(TTF)_4[7a]_3$ and the material exhibited a conductivity of 9×10^{-6} S cm^{-1}.[24]

Table 5. Selected bond distances (Å) of $[Bu_4N]_2[7a]$ and $[Bu_4N]_2[9]$.

Parameter	$[Bu_4N]_2[7a]^a$	$[Bu_4N]_2[9]^b$
Ni–S1	2.190(1)	2.1805(5)
Ni–S2	2.185(1)	2.1952(5)
S1–C1	1.735(4)	1.741(2)
S2–C2	1.747(4)	1.741(2)
C1–C2	1.436(5)	1.427(3)
C1–C4	1.387(5)	1.372(3)
C2–C3	1.378(6)	1.376(3)
S3–C3	1.756(4)	1.749(2)
S3–C4	1.740(4)	1.748(2)
^{a}Reference 22. ^{b}Reference 24.		

Conductive materials were also produced via oxidative electro-crystallization from solutions of [Et$_4$N][**10a**] and [(PPh$_3$)$_2$N][**10c**] in acetonitrile, giving black microcrystalline powders. Elemental analyses of the resulting powders indicated relatively neutral materials, with very little cation (Et$_4$N$^+$ or (PPh$_3$)$_2$N$^+$) remaining in the solids. The conductivities of the materials were found to be 2×10^{-7} S cm^{-1} for the electrochemically oxidized product of **10a** and 7×10^{-8} S cm^{-1} for that of **10c**.[24]

Other Thiophene-Functionalized Metal Dithiolene Complexes

Synthesis. The last class of thiophene-containing metal dithiolene complexes includes those species with pendent thiophene groups attached to a classical metal dithiolene core (Scheme 6). The earliest example of this group was reported by Freyer with the generation of the neutral complex **12** in 1984.[25] Unfortunately, no synthetic details were included in this work, but it does seems implied that **12** was prepared using the previous methods of Schrauzer and Mayweg.[26] This chemistry was later revisited by Pickup and coworkers in 2001, who prepared the intermediate Ni dithiolene salt [Et$_4$N][**11**] in a manner similar to the previous salts discussed above.[27,28] Oxidation of [Et$_4$N][**11**] in acetonitrile then allowed the isolation of the neutral species **12**. In addition, it was found that utilization of the preligated Ni(dppe)Cl$_2$ complex in place of NiCl$_2$ allowed the isolation of the analogous, mixed ligand dithiolene complex **13**. To further extend this work, the related salt [Et$_4$N][**14**] was also prepared via similar methods and oxidized to the give the neutral complex **15**.[28]

More recently, an analogous salt of **11** and the neutral **12** were prepared by Robertson, Murphy, and coworkers using a slightly different synthetic route (Scheme 7).[29] Here, the ligand precursor 4,5-bis(2-thienyl)-1,3-dithiol-2-one was prepared through a route previously developed by Murphy and coworkers for the production of thiophene-functionalized TTF species.[30] Production of the anionic and neutral Ni dithiolenes were then accomplished via similar methods to those used by Pickup and coworkers above. Using analogous methods, the 3-thienyl-functionalized isomers [Bu$_4$N][**16**] and **17** were also prepared starting from 3-thiophenecarboxaldehyde.[29]

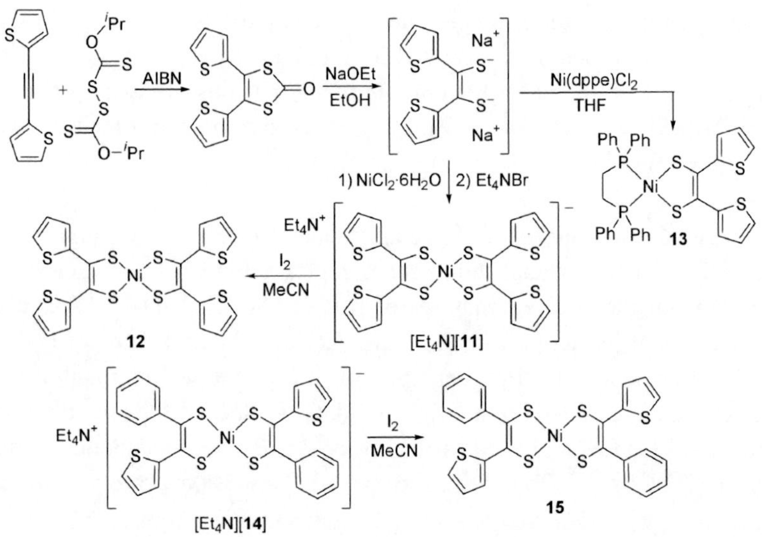

Scheme 6. Synthesis of metal dithiolenes containing pendent thiophenes.

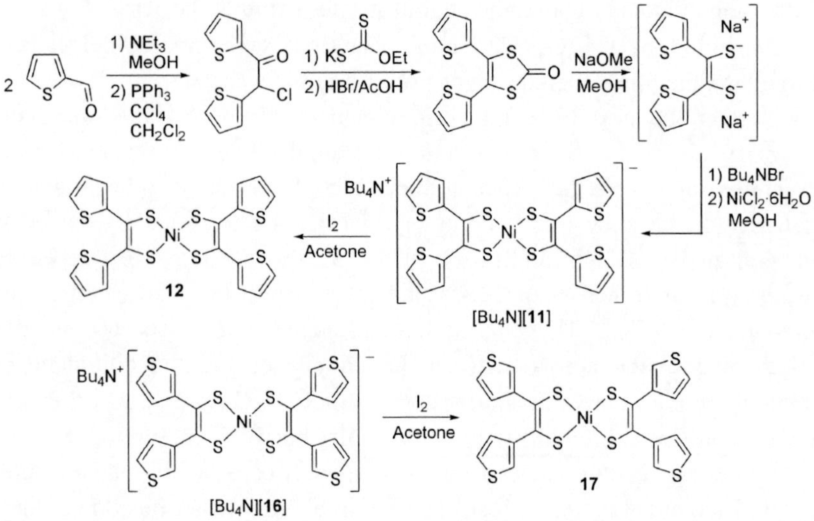

Scheme 7. Alternate synthesis of metal dithiolenes containing pendent thiophenes.

In addition to the preparation of the neutral species **12** and **17** via I_2 oxidation of the corresponding anions, it was found that **17** could be prepared in higher purity by simple recrystallization of [Bu$_4$N][**16**] from acetone and ethanol. Here, oxidation to the neutral complex takes place during the recrystallization process as observed by a color change from the deep red of the initial anion to the pale green/brown of **17**. The observed ease of oxidation of [Bu$_4$N][**16**] is consistent with the lower positive potential measured for its −1/0 redox couple in comparison to [Bu$_4$N][**11**].[29]

X-ray Crystallography. Several species within this final class of thiophene-containing metal dithiolene complexes have been studied by X-ray crystallography, with the structures of [Et$_4$N][**11**], [Et$_4$N][**14**], **12**, and **17** all reported.[28,29] Of particular interest is the fact that the structures of [Et$_4$N][**11**] and **12** represent the first example of a thiophene-functionalized metal dithiolene characterized in multiple oxidation states. Ellipsoid plots of the complexes **11** and **12** are shown in Figure 12 and selected bond lengths are given in Table 6.[28,29] Both complexes exhibit inversion centers at the nickel metal center, with observed Ni–S bond lengths ranging from 2.125 to 2.151 Å. The anion exhibits longer Ni–S bonds than the neutral complex, which agrees with the previously discussed trend that the M–S bond lengths of metal dithiolenes roughly correlate in an inverse fashion with the complex charge.[1] Within the dithiolate ligand, the neutral species exhibits shorter C–S bonds but an elongation of the C1–C2 bond, consistent with greater delocalization throughout the 5-membered ring of the Ni-dithiolate chelate.

Neither **11** nor **12** exhibit significant distortion of the coordination geometry about the Ni center and are completely square planar. In both structures, however, the pendent thiophenes show distinct deviations from planarity. In the case of **11**, both sets of rings are significantly canted with dihedral angles of 30.5° and 66.7°,[28] while **12** shows one set of thiophenes to be near-planar (dihedral of 12.4°) and the other to be nearly perpendicular (dihedral of 79.3°).[29] These large dihedral angles could also account for the lengthening of the bonds between the thiophenes and the dithiolene core. While these bonds are still shorter than a simple C–C single bond, they are all longer than that exhibited by the interannular bond of 2,2'-bithiophenes (1.45 Å),[20] thus suggesting reduced conjugation between the thiophenes and the metal dithiolene. Lastly, both structures exhibit significant positional disorder within the pendent thiophenes, which results in partial occupation of the sulfur sites by carbon.

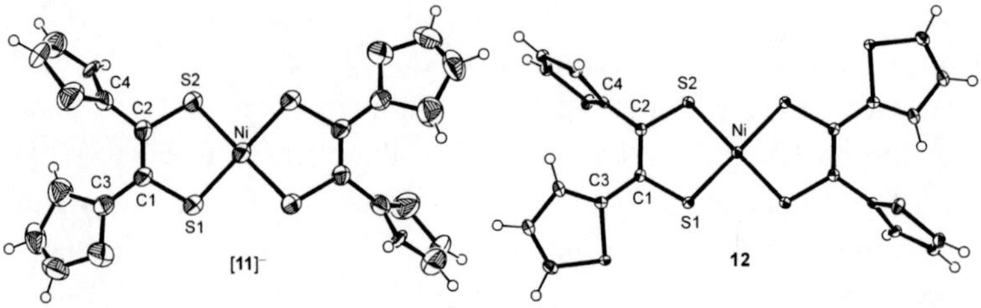

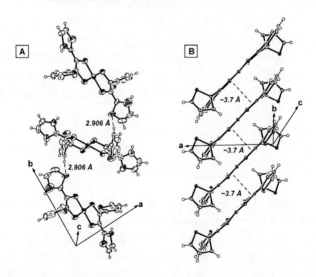

Figure 12. Ellipsoid plots of the anion **11** and its neutral analogue **12** at the 50% probability level.

Table 6. Selected bond distances (Å) of [Et₄N][11] and 12.

Parameter	[Et₄N][11][a]	12[b]
Ni–S1	2.151(1)	2.1303(7)
Ni–S2	2.139(1)	2.1250(7)
S1–C1	1.740(4)	1.715(3)
S2–C2	1.734(4)	1.706(2)
C1–C2	1.364(6)	1.400(3)
C1–C3	1.474(5)	1.464(3)
C2–C4	1.481(6)	1.476(4)
[a]Reference 28. [b]Reference 29.		

Figure 13. Short contacts between anions in [Et₄N][**11**] (A) and π-stacked columns of **12** (B).

The nonplanar nature of **11** hinders intermolecular contacts between the metal dithiolene complexes. Thus, there are few intermolecular interactions between the anions within [Et$_4$N][**11**], although short C–H···S contacts of 2.906 Å are observed between the pendent thiophenes of neighboring anions (Figure 13A).[28] In contrast, the more planar structure of **12** allows significant π stacking of the complexes to form columns with a face-to-face distance of ~3.7 Å (Figure 13B).[29]

The remaining two related structures of [Et$_4$N][**14**] and **17** exhibit similar features. The anion **14** adopts a structure similar to **12** with the thiophenes near planar (dihedral of 9.0°) and the phenyl rings near perpendicular (dihedral of 79.6°).[28] The bond between the thiophenes and the dithiolene core is 1.472 Å, roughly the same as the two previous structures. No significant intermolecular interactions were observed between the metal thiophenedithiolene anions.

Complex **17** adopts a structure with all pendent thiophene rings significantly canted with dihedral angles of 31.5, 44.2, 47.7, and 52.7°.[28] As with **12**, the molecules of **17** exhibit some π stacking, although here the complexes have less direct overlap and are slipped along the short molecular axis. The face-to-face distance between complexes in this slip-stacked arrangement is a bit larger (~3.8 Å) than that seen in **12**, indicative of a weaker π interaction.

Solid State Properties. Most likely as a result of the limited intermolecular interactions of these complexes in the solid state, little investigation of their solid-state properties have been reported. However, the electrical conductivity of pressed pellets of **17** have been measured via four-point probe to give a room temperature value of 10^{-5} S cm^{-1}.[29] This relatively low value is likely a result of the lack of significantly strong π-stacking as observed in the X-ray structure.

Metal Dithiolene-Based Conjugated Polymers

A main focus of preparing metal dithiolenes functionalized with thiophene units has been for their incorporation into conjugated organic materials. On first glance, it would seem that the combination of metal dithiolenes and conjugated polymers could produce attractive new materials for use in such applications as field effect transistors and NIR optical materials. However, several challenges remain before these materials can be applied to useful devices. First, synthetic access to a wide variety of polymer structures remains a difficult task, especially structures which would yield solution processable polymers. To date, the only polythiophene/dithiolene polymers synthesized have been insoluble materials produced by electropolymerizaton.[18,22,27-29] Secondly, although there have been a few studies which incorporate dithiolenes into conjugated polymers, understanding structure-property relationships remains a challenge due to the few structures that are available and the limited methods of analysis for the resulting insoluble materials. For instance, it seems that in the studies presented in this section, polymers containing dithiolenes as pendant groups give electrochemical and optical response resembling both polythiophenes and dithiolenes, but polymers containing dithiolenes in the main chain of conjugation seem to give materials which indicate that the dithiolene unit does not mediate conjugation through the chains. So far, little explanation for this behavior has been proposed. Overall, the performance of the polymers prepared so far has been somewhat disappointing as far as material properties such as conductivity, and as such these materials have not been incorporated into working devices to date. The structures of the polymers reviewed in this section are shown in Chart 3.

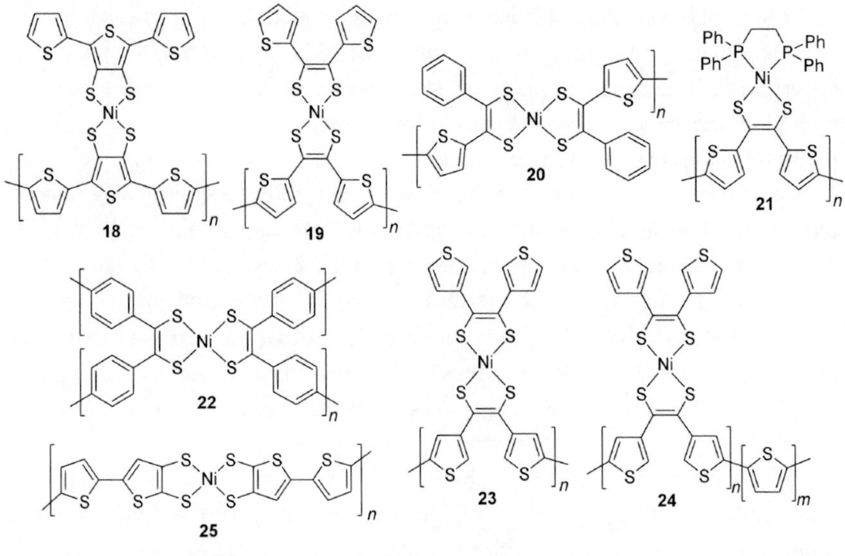

Chart 3. Structures of dithiolene-containing conjugated polymers.

The synthesis of Polymer **18** by anodic polymerization of [Bu$_4$N][**7a**] on ITO-coated glass or glassy carbon electrodes was reported by Skabrara and coworkers,[22] and the optical and electrochemical properties of the resulting polymer films were studied. This polymer displayed polythiophene-like electrochemical behavior at positive potentials and metal dithiolene-like reductions at negative potentials. The polymer also exhibited a broad UV-visible-NIR absorption out to 1100 nm, which caused the authors to propose its use in light harvesting components in photovoltaic devices, although to date this application has not been investigated.

In another work, polymer **19** was synthesized by anodic polymerization of neutral **12** on platinum and ITO electrodes.[27] Polymer **19** gave three distinct redox waves, and the band gap of the material was about 0.45 eV. It's disappointingly low conductivity (10^{-6} S cm^{-1}) in the 0/–1 mixed valence state and only slightly higher (10^{-4} S cm^{-1}) conductivity in an oxidized state as measured by impedance spectroscopy caused the authors to hypothesize that cross-linking was giving materials with fairly short conjugation lengths. In order to gain insight into the behavior of **19** and produce materials that would form long chains with extended conjugation, the authors synthesized polymers **20-22**.[28] Polymers **20** and **21** were synthesized by anodic polymerization of **15** and **13**, respectively, and **22** was synthesized by cathodic polymerization of its tetrabrominated monomer in the presence of Ni(bpy)$_3$(ClO$_4$)$_2$ (bpy = 2,2'-bipyridine). In contrast to polymer **19**, polymer **20** did not exhibit chemically reversible redox behavior, and thus it was shown that the nickel center was not mediating conjugation between repeat units. The electrochemical behavior of **21** was similar to polythienylenevinylene,[31] exhibiting a chemically reversible oxidation wave and no reduction waves indicative of metal dithiolene units. The electrochemical behavior of the analogous **22** was also similar to that of polyphenylenevinylene, with partially reversible reduction waves occurring around –1.7 V vs. SSCE. The authors concluded that the facile oxidation of **19** and **21** was a result of mostly thienylenevinylene-type linkages in the polymer backbone, although the possibility of significant cross-linking in **19** could not be ruled out.

In a subsequent study by Robertson, Murphy, and coworkers, polymer **23** was synthesized by anodic polymerization on platinum electrodes and compared to polymer **19** in order to determine the effect of the position of the linkage of the dithiolene to thiophene moieties.[29] In this study, the researchers used the salts [Bu$_4$N][**11**] and [Bu$_4$N][**16**] in MeCN to prepare the respective polymers **19** and **23**. Surprisingly, the polymerization of the anion **11** resulted in films exhibiting different electrochemical response than that observed for **19** prepared from the polymerization of **12** in CH_2Cl_2 as described in the previous study.[27,28] In this case, the electrochemical behavior of **19** was similar to the behavior of polythiophenes, but showed no evidence of intact dithiolene units and no significant reductions with any reversibility were evident at potentials more negative than 0.6 V vs. SCE. The researchers attributed this discrepancy to the use of the different solvents for the polymerization rather than the neutral or anionic nature of the precursor complex. Polymer **23**, on the other hand, was deposited as an insulating film on the surface of the electrode from an acetonitrile solution. Polymer **23** was dissolved in dichloromethane to give an almost identical UV-vis-NIR spectrum to that of the monomer, indicating intact dithiolene units. One explanation given by the authors for the insulating nature of **23** was that the X-ray crystal structures of monomers **12** and **17** show significant twisting of the thiophenes with respect to the nickel dithiolene core, due to steric interactions between adjacent thiophene rings. This would cause limited or no conjugation between repeat units. Interestingly, it was shown that two different materials could be deposited on the electrodes, depending on the potential that was used. At +1.39 V vs. SCE, an insulating polymeric film was deposited, as mentioned above. This film was then dissolved in CH_2Cl_2, and it showed a NIR absorption peak indicative of intact metal dithiolene units. At +0.29 V vs. SCE, a conducting film of a molecular neutral complex **17** was deposited, and upon reduction at −0.75 V the film redissolved into the MeCN solution.

The authors also showed that [Bu$_4$N][**16**] could be anodically copolymerized with thiophene in a 1:16 feed ratio at +1.79 V vs. SCE to give films of significant conductivity, although lower than polythiophene. While the films had significantly less solubility in CH_2Cl_2 than polymer **23**, the soluble portion of the films also showed the NIR absorption indicative of intact dithiolene units, along with an increase in relative absorption intensity from 230-400 nm. This was an indication that short thiophene oligomers with incorporated **16** were able to dissolve into CH_2Cl_2, but it was hypothesized that the insoluble portion contained the longer polythiophene chains which would significantly absorb out to 500 nm.[32] This important result showed that dithiolenes could be successfully electropolymerized with other anodically polymerizable monomers to give hybrid polymers with properties characteristic of both homopolymers.

Amb and Rasmussen attempted to synthesize linear polymer **25** by two different electrochemical methods, anodic polymerization of [Bu$_4$N][**4a**] in a number of solvents, and cathodic polymerization of [Bu$_4$N][**4e**] in the presence of [Ni(bpy)$_3$][PF$_6$]$_2$ in a 0.1 M Bu$_4$NPF$_6$ dichloromethane solution.[18] The cyclic voltammagram of [Bu$_4$N][**4a**] showed two irreversible oxidation waves, one corresponding to a −1 to neutral or mixed-valence charge state around −0.15 V vs. Fc/Fc$^+$ in which there was adsorption onto the corresponding electrode, and another at higher potential (+0.70 V vs. Fc/Fc$^+$) corresponding to further oxidation of the adsorbed species. Holding electrode potentials at the most positive peak resulted in deposition of a black film in 0.1 M Bu$_4$NPF$_6$ solutions of DMF, MeCN, and CH$_2$Cl$_2$, but the films did not display significant electroactivity. Conducting films of **4a** *could* be grown by holding electrode potentials at around 0 V vs. Fc/Fc$^+$ for extended periods of time (>3 h) in solutions of [Bu$_4$N][**4a**], however. Surprisingly, the film did not desorb from the electrode upon repeated reduction/cycling, possibly indicating a highly ordered and kinetically stable material.

As it was hypothesized that the oxidation could be destroying the dithiolene moiety, Yamamoto polymerization was employed to ensure that the electrochemical conditions would not irreversibly damage the metal dithiolene unit. Reduction of the [Ni(bpy)$_3$][PF$_6$]$_2$ occurred at potentials which gave reversible electrochemical behavior of [Bu$_4$N][**4e**], and resulted in the deposition of a black film which again did not exhibit significant electro-activity. However, it did give an NIR absorption indicative of intact metal dithiolene units. Additionally, after deposition of the film little electrochemical response was evident in the monomer solution, indicating that the film was insulating. Interestingly, the absorption maximum of a film deposited on ITO was blue shifted by ~110 nm from that of the monomer, but the onset of absorption was red-shifted to ~1750 nm, thus giving a band gap for the deposited film of ~0.7 eV (Figure 14).

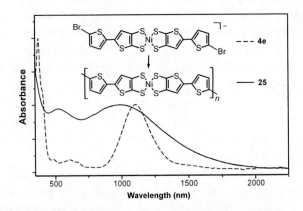

Figure 14. UV-Visible-NIR absorption spectra of [Bu$_4$N][**4e**] and **25**.

The incorporation of metal dithiolene moieties into conjugated polymers has been a difficult and sometimes frustrating undertaking. Although the idea seems attractive, it has been shown here that these polymers are synthetically difficult to achieve, and often give disappointing properties. A major point of concern and still an unsolved problem is the lack of conductivity of polymers with metal dithiolenes incorporated as part of the main conjugation path in a polymer backbone. A possible reason for this could be that in their neutral charge state, dithiolenes with appended aryl groups have little to no contribution from the nickel center to filled frontier molecular orbitals,[31] thus creating a conjugation break when it is possible for thiophene-containing conjugated polymers to be conducting in their oxidized states. Indeed, more theoretical work is necessary to understand this problem. The other main limitation to the application of these polymers is the lack of synthetic methods necessary to incorporate these into solution-processable materials. So far, the most attractive and simple approach for the future use of these types of materials in real applications is probably the conjugated polymer/metal dithiolene hybrid approach first presented by Robertson, Murphy, and coworkers.[29]

ACKNOWLEDGEMENTS

Crystallographic data was retrieved through the Cambridge Crystallographic Data Centre (CCDC)[34] and all crystallographic figures were generated from the cif files using the Ortep3 software package.[35]

REFERENCES

[1] *Dithiolene Chemistry: Synthesis, Properties, and Applications*, EI. Tiefel, KD. Karlin, Eds; *Progress in Inorganic Chemistry*, Vol. 52, John Wiley & Sons: Hoboken, 2004.

[2] Robertson, N; Cronin, L. *Coord. Chem. Rev.*, 2002, 227, 93.

[3] Kato, R. *Chem. Rev.*, 2004, 104, 5319.

[4] Gol'dfarb, YaL; Kalik, MA. *Khimiya Geterotsiklicheskikh Soedinenii* 1968, 788, *Chem Abstracts* 70, 96512.

[5] Belo, D; Alves, H; Lopes, EB; Gama, V; Henriques, RT; Duarte, MT; Almeida, M; Perez-Benitez, A; Rovira, C; Veciana, J. *Synth. Met*, 2001,

120, 699.

[6]   Belo, D; Alves, H; Rabaca, S; Pereira, LC; Duarte, MT; Gama, V; Henriques, RT; Almeida, M; Ribera, E; Rovira, C; Veciana, J. *Eur. J. Inorg. Chem.*, 2001, 3127.

[7]   Belo, D; Alves, H; Lopes, EB; Duarte, MT; Gama, V; Henriques, RT; Almeida, M; Pérez-Benítez, A; Rovira, C; Veciana, J. *Chem. Eur. J,* 2001, 7, 511.

[8]   Belo, D; Figueira, MJ; Mendonça, J; Santos, IC; Almeida, M; Henriques, RT; Duarte, T; Rovira, C; Veciana, J. *Eur. J. Inorg. Chem.,* 2005, 3337.

[9]   Belo, D; Figueira, MJ; Nunes, JPM; Santos, IC; Pereira, LC; Almeida, M; Rovira, C. *J. Mater. Chem.*, 2006, 2746.

[10] Belo, D; Mendonça, J; Santos, IC; Pereira, LCJ; Almeida, M; Novoa, J. J; Rovira, C; Veciana, J; Gama, V. *Eur. J. Inorg. Chem.*, 2008, 5327.

[11] Katritzky, AR; Pozharskii, AF. *Handbook of Heterocyclic Chemistry*, 2nd ed; Pergamon: New York, 2000, 61.

[12] Ogawa, K; Rasmussen, SC. *J. Org. Chem.*, 2003, 68, 2921.

[13] Bondi, A. *J. Phys. Chem.*, 1964, 68, 441.

[14] Desiraju, GR. *Acc. Chem. Res.*, 2002, 35, 565.

[15] Claessens, CG; Stoddart, JF. *J. Phys. Org. Chem.*, 1997, 10, 254.

[16] Amb, CM; Rasmussen, SC. *Polym. Prepr*, 2007, 48, 128.

[17] Amb, CM; Rasmussen, SC. *Synth. Met*, 2009, 159, 2390.

[18] Amb, CM; Heth, CL; Evenson, SJ; Pokhodnya, K; Rasmussen, SC. submitted to *Inorg. Chem.*

[19] Amb, CM; Rasmussen, SC. *Eur. J. Org. Chem.*, 2008, 801.

[20] Barbarella, G; Zambianchi, M; Antolini, L; Folli, U; Goldoni, F; Larossi, D; Schenetti, L; Bongini, A. *J. Chem. Soc. Perkin Trans*, 2 1995, 1869.

[21] Chiang, LY; Shu, P; Holt, D; Cowan, D. *J. Org. Chem.*, 1983, 48, 4713.

[22] Pozo-Gonzalo, C; Berridge, R; Skabara, PJ; Cerrada, E; Laguna, M; Coles, SJ; Hursthouse, MB. *Chem. Commun*, 2002, 2408.

[23] Pozo-Gonzalo, C; Khan, T; McDouall, JJW; Skabara, PJ; Roberts, DM; Light, ME; Coles, SJ; Hursthouse, MB; Neugebauer, H; Cravinod, A; Sariciftci, NS. *J. Mater. Chem.*, 2002, 12, 500.

[24] Skabara, PJ; Pozo-Gonzalo, C; Miazza, NL; Laguna, M; Cerrada, E; Luquin, A; González, B; Coles, SJ; Hursthouse, MB; Harrington, RW; Clegg, W. *Dalton Trans*, 2008, 3070.

[25] Freyer, W. *Z. Chem.*, 1984, 24, 32.

[26] Schrauzer, GN; Mayweg, VP. *J. Am. Chem. Soc.*, 1965, 87, 1483.

[27] Kean, CL; Pickup, PG. *Chem. Commun*, 2001, 815.

[28] Kean, CL; Miller, DO; Pickup, PG. *J. Mater. Chem.*, 2002, 12, 2949.
[29] Anjos, T; Roberts-Bleming, SJ; Charlton, A; Robertson, N; Mount, AR; Coles, SJ; Hursthouse, MB; Kalaji, M; Murphy, PJ. *J. Mater. Chem.*, 2008, 18, 475.
[30] Charlton, A; Underhill, AE; Williams, G; Kalaji, M; Murphy, PJ; Malik, KMA; Hursthouse, MB. *J. Org. Chem.*, 1997, 62, 3098.
[31] Onodat, M; Iwasa, T; Kawai, T; Yoshino, K. *Phys. D: Appl. Phys.*, 1991, 24, 2076.
[32] J. Roncali, *Chem. Rev.*, 1992, 92, 711.
[33] Ray, K; Weyhermuller, T; Neese, F; Wieghardt, K. *Inorg. Chem.*, 2005, 44, 5345.
[34] Allen, FH. *Acta Cryst*, 2002, B58, 380.
[35] Farrugia, LJ. *J. Appl. Cryst*, 1997, 30, 565.

Reviewed by Dr. Neil Robertson, School of Chemistry, University of Edinburgh.

In: Chemical Crystallography
Editors: Bryan L. Connelly, pp. 103-130 © 2010 Nova Science Publishers, Inc.

ISBN: 978-1-60876-281-1

Chapter 3

CRYSTAL CHEMISTRY OF AN ATROPISOMER: CONFORMATION, CHIRALITY, AROMATICITY AND INTERMOLECULAR INTERACTIONS OF DIPHENYLGUANIDINE

Manuela Ramos Silva, Pedro S. Pereira Silva,*
Ana Matos Beja and José António Paixão
Cemdrx, Physics Department, University of Coimbra, P-3004-516
Coimbra, Portugal.

ABSTRACT

This paper reviews the crystal chemistry of 25 diphenylguanidine/diphenylguanidinium compounds. Diphenylguanidine is an atropisomer and several conformations have been isolated in the solid state. Such conformations are investigated and the conformation description, chirality, aromaticity of the flexible molecule are systematized in this paper. The dipolar moment and octupolar character are probed. Intermolecular interactions are classified. Two new salts are reported: N,N'-Diphenylguanidinium nicotinate hydrate and N,N'-Diphenylguanidinium 5-nitrouracilate dihydrate. The former crystallizes in a chiral space group, with the phenyl rings of the cation oriented like

---

* Corresponding author: Email: manuela@pollux.fis.uc.pt.

the blades of a propeller. The latter crystallizes in the centrosymmetric, triclinic space group P-1, and the cation exhibits and *anti-anti* conformation.

Keywords: atropisomer; polarizability; X-ray diffraction.

INTRODUCTION

Diphenylguanidine also known as melaniline, is used as a cure accelerator in the rubber industry. It is marked under the trade name of "Vulkazit". Certain N,N′-diarylguanidines are potent ligands for the N-methyl-D-aspartate/PCP(Phencyclidine) receptor and have neuroprotective properties against glutamate-induced neuronal cell death [1]. N,N′-di-ortho-tolylguanidine and its congeners are selective ligands for the haloperidol-sensitive σ receptor [2,3]. As such, di-substituted guanidine compounds are of considerable interest in pharmaceutical applications, as neuroleptic and antipsychotic drugs.

From the point of view of their physical properties, guanidine compounds are potentially interesting for non-linear applications [4]. Non-linear materials are those in which the dielectric polarization responds nonlinearly to the electric field of the incident light, producing different frequencies. Such materials are used associated with lasers enabling the expansion of the lasers limited spectral regime. Guanidinium moieties due to their D_{3h} symmetry guarantee a strong octupolar moment and a small dipole moment thus circumventing the tendency of large dipolar molecules to assemble in centrosymmetric space groups (that extinguish the non-linear optical responses to stimuli).

Di-substituted neutral guanidines can exist in two tautomeric forms, the imino and amino form (Scheme 1). The imino form is favoured in monoalkylguanidines as L-arginine [5] and the amino form is favoured in monoarylguanidines [6]. The latter tautomeric preference is seen in the solid state structures containing neutral diphenylguanidine [7-8].

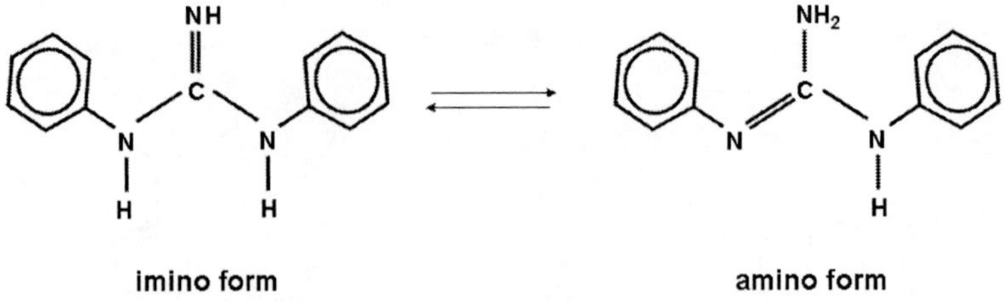

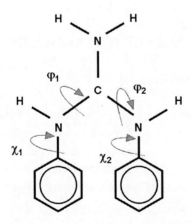

imino form   amino form

Scheme 1. Tautomeric forms of neutral diphenylguanidine.

The diphenylguanidine molecule or the diphenylguanidinium positive ion show several conformational isomers. The phenyl rings can rotate around the C-N single bonds and three types of main conformations are observed in the solid state, *syn-syn*, *syn-anti* and *anti-anti* (Schemes 2 and 3).

The N,N'-diphenylguanidinium (dpg+) has been found to adopt different conformations both in aqueous solutions [9] and in several salts that are being reviewed in this paper. The conformation of dpg+ is very sensitive to the counter-ion, and this effect has been the subject of *ab-initio* quantum mechanical and molecular mechanics calculations [10]. Stabilization of a particular conformation depends critically on intermolecular interactions with the solvent, since the energetic cost of rotation of the phenyl rings is much lower than typical solvation energies.

Scheme 2. Rotation possibilities for the diphenylguanidinium cation and definition of torsion angles φ_1, φ_2, χ_1, χ_2.

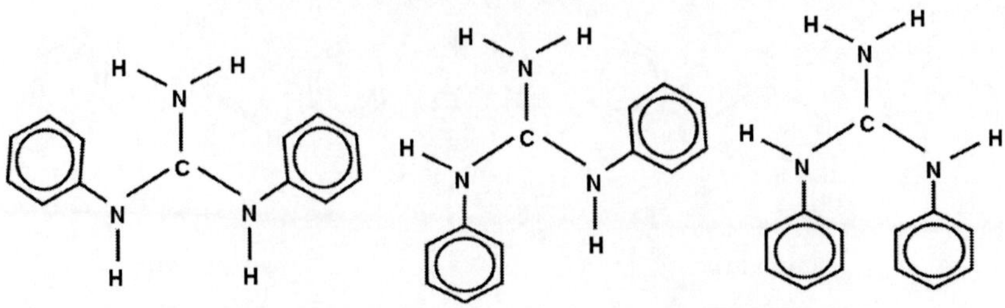

Scheme 3. *Syn-syn, anti-syn* and *anti-anti* conformation of the diphenylguanidinium cation.

The *syn-anti* conformation is particularly suited for the orientation of the phenyl rings as the blades of a propeller. This induces chirality in an otherwise achiral molecule: atropisomerism, conformational stereoisomerism, is therefore observed in this family of compounds.

The flexible conformation of dpg/dpg+ is analysed in this paper by reviewing the structural data of 23 already reported structures and by presenting two new ones: N,N'-Diphenylguanidinium nicotinate hydrate and N,N'-diphenylguanidinium 5-nitrouracilate dihydrate. The reported data were extracted from CSD database release 5.29 [11] and only CIF files containing well ordered dpg/dpg+ moieties were used. When multiples studies exist, only the one with better quality factors was chosen.

EXPERIMENTAL SECTION

Preparation of Compound 1

0.2514g (1 mmol) of nicotinic acid (Aldrich, 98%) were dissolved in 25 ml of water and 0.4354 mg (2 mmol) of diphenylguanidine (Aldrich 97%) were dissolved in 75 ml of ethanol. The acid solution was slowly added to the ethanolic solution of dpg. The resultant solution was stirred for 30 min at 40°C and left to evaporate under ambient conditions. After a period of a few weeks, small prism-like transparent single crystals with appropriate quality for X-ray measurements were grown.

Preparation of Compound 2

0.1603g (1 mmol) of 5-nitrouracil (Aldrich, 98%) were dissolved in 40 ml of water and 0.2178g (1 mmol) of diphenylguanidine (Aldrich 97%) were dissolved in 50 ml of ethanol. The acid solution was slowly added to the basic solution. The resultant solution was heated to the boiling point and was then left to evaporate at room temperature and pressure. After 3 weeks, small transparent single crystals were obtained.

Crystal Structure Determination

For diphenylguanidinium nicotinate hydrate, a colourless crystal having approximate dimensions of 0.35 mm x0.20 mm x 0.10 mm was glued at the top of a glass fiber and mounted on a APEXII difractometer. Diffraction data were collected at room temperature [293(2) K] using graphite monochromated Mo Kα radiation(λ= 0.71073 Å). The unit cell parameters were determined by least-squares refinement of diffractometer angles (2.56 °<θ<25.18°) for 8578 reflections.

An absorption correction was applied which resulted in transmission factors ranging from 0.891 to 0.991. Reflections with 2θ ≤ 50° were used for structure solution and refinement.

The structure was solved by direct methods using SHELXS-97 [12] and refined anisotropically (non-H atoms) by full-matrix least-squares on F^2 using the SHELXL-97 [12] program . All the hydrogen atoms were placed at calculated positions and allowed to ride on their parent atoms using SHELXL-97 defaults, with exception to the water hydrogen atoms that were located in a Fourier difference map (freely refined atomic coordinates). The final least-squares cycle was based on 2940 observed reflections [I>2σ(I)] and 241 variable parameters, converged to R = 0.0375 and wR = 0.0996. In the absence of significant anomalous scattering effects, Friedel pairs were merged in the final refinement. The crystallographic details and selected interatomic distances and angles are given in Tables 1-2.

For diphenylguanidinium 5-nitrouracilate dihydrate, a slightly yellowish crystal having approximate dimensions of 0.36 mm x 0.19 mm x 0.03mm was placed at the top of a glass fiber and mounted on a APEXII difractometer. Diffraction data were collected at room temperature 293(2) K using graphite monochromated Mo Kα (λ= 0.71073 Å). The unit cell parameters were determined by least-squares refinement of diffractometer angles (3.03 °<θ<22.62°) for 4093 reflections.

An absorption correction was applied which resulted in transmission factors ranging from 0.914 to 0.997. Reflections with $2\theta \leq 50°$ were used for structure solution and refinement.

The structure was solved by direct methods using SHELXS-97 [12] and refined anisotropically (non-H atoms) by full-matrix least-squares on F^2 using the SHELXL-97 [12] program . All the hydrogen atoms were placed at calculated positions and allowed to ride on their parent atoms using SHELXL-97 defaults, with exception to the water hydrogen that were located in a Fourier difference map (freely refined atomic coordinates). The final least-squares cycle was based on 2602 observed reflections [I>2σ(I)] and 274 variable parameters, converged to R = 0.0454 and wR = 0.1266. The crystallographic details and selected interatomic distances and angles are given in Tables 4-5.

RESULTS

N,N'-Diphenylguanidinium   nicotinate   crystallizes   in   a   non-centrosymmetric space group, $P2_12_12_1$ with four positive ions and four negative ions per unit cell. The bond lengths C1-N1 [1.338(2) Å] and C1-N2 [1.338(2) Å] of the guanidinium group are close to the standard value of a delocalized C=N bond [1.339(5) Å] while the bond length C1-N3 [1.315(2) Å] is somewhat shorter. These distances are comparable with the average values relevant to guanidinium cations such as 1.321 and 1.328 Å for unsubstituted and substituted species [13]. The phenyl rings are practically flat and have the usual geometry: mean values of the C-C bond length are 1.380(5) for ring C2-C7 and 1.381(4) for C8-C13, the average absolute torsion angles are 1.0(2)° and 0.9(2)° for rings C2-C7 and C8-C13, respectively. The phenyl rings are oriented in an *anti-syn* conformation (Figure. 1).

The angle between the least-squares planes of the two phenyl rings is 85.66(11)° and the angles between the central plane and the phenyl rings are 65.74(11)° and 54.30(11)° respectively for rings C2-C7 and C8-C13. The negative ion, the nicotinic moiety, has the carboxylic group slightly rotated from the aromatic ring plane as seen by the torsion angle, C16-C15-C14-O1 152.00(17)°. The C-O distances are characteristic of a deprotonated carboxylic group with a delocalized character, O1-C14 1.254(2) and O2-C14 1.253(2) Å.

The C-N distances within the ring are C16-N4 1.346(2)Å and C17-N4 1.332(2) Å.

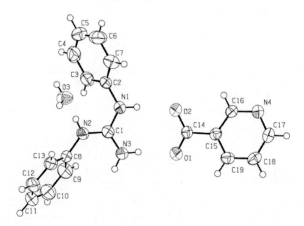

Figure 1. ORTEP diagram of diphenylguanidinium nicotinate hydrate. The thermal ellipsoids were drawn at the 50% probability level.

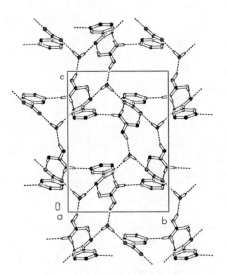

Figure 2. Packing of the molecules in the diphenylguanidinium nicotinate hydrate crystal structure. The dashed lines represent hydrogen bonds and some atoms were omitted for clarity.

Table 1. Summary of crystallographic results for diphenylguanidinium nicotinate.

Temperature (K)	293(2)
Empirical formula	$C_{19}H_{20}N_4O_3$
Formula weight	352.39
Wavelength (Å)	0.71073
Crystal system	orthorhombic
Space group	$P2_12_12_1$
a (Å)	10.1356(2)
b (Å)	11.6112(2)
c (Å)	15.9465(3)
Volume (Å^3)	1876.69(6)
Z	4
Calculated density (g/cm^3)	1.247
Absorption coefficient (mm^{-1})	0.087
F(000)	744
Crystal size (mm^3)	0.35 x 0.20 x 0.10
θ range for data collection (deg.)	2.17 - 25.0
Index ranges	$-13 \leq h \leq 13, -15 \leq k \leq 16, -20 \leq l \leq 20$
Reflections collected/unique	63396 / 2940
Completeness to θ=50°	100%
Transmission factors (min/max)	0.899/0.999
Data/restraints/parameters	2940/0/241
Goodness–of–fit on F^2	1.030
Final R indices [I> 2σ(I)]	0.0375 /0.0574
R indices (all data)	0.0932 /0.0996
Largest diff. peak and hole (e Å^{-3})	–0.179 /0.165
CCDC Number	CCDC 724892

There is a three-dimensional hydrogen bond network linking anions, cations and water molecules thus re-enforcing the crystal cohesion (Table 3). The water molecules form *zig-zag* chains together with the nicotinate ions that run along the *b*-axis (Figure. 2). Such chains are then interconnected by H-bonds though the guanidinium fragments.

Table 2. Selected bond lengths (Å) and angles (°) for diphenylguanidinium nicotinate hydrate.

N1 – C2	1.427(2)
N2 – C8	1.421(2)
C1 – N1 – C2	124.34(14)
C1–N2–C8	125.76(14)
N3 – C1 – N1	118.97(16)
N3 – C1 – N2	121.73(16)
N1 – C1 – N2	119.30(16)
C17 – N4 – C16	117.19(16)

Table 3. H-bonding geometry for diphenylguanidinium nicotinate hydrate (Å, °).

Donor-H....Acceptor	D - H	H...A	D...A	D - H...A
N1-H1 ... O2	0.86	1.98	2.838(2)	173
N2-H2 ... O3	0.86	2.02	2.7958(19)	150
N3-H3A ... O1	0.86	1.93	2.771(2)	164
N3-H3B ... N4^i	0.86	2.11	2.852(2)	144
O3-H31 ... O1^{ii}	0.88(3)	1.98(3)	2.855(2)	179(4)
O3-H32 ... O2^{iii}	0.79(3)	2.07(3)	2.855(2)	171(3)

i: 2-x,-1/2+y,3/2-z, *ii*: 3/2-x,-y,1/2+z, *iii*: -1/2+x,1/2-y,2-z .

There are also some C-H …π intermolecular interactions. Such interactions use the π cloud of the aromatic rings as acceptors for the hydrogen protons.

The compound N,N'-diphenylguanidinium 5-nitrouracilate dihydrate cristallizes in a centrosymmetric space group with one anion, one cation and two water molecules per asymmetric unit cell (Figure. 3). The inversion centre that relates the ions in the unit cell obliterates the non-linear optical response of this compound. Other 5-nitrouracil compounds permit efficient blue-light generation from a pulsed laser source [14].

Table 4. Summary of crystallographic results for diphenylguanidinium 5-nitrouracilate dehydrate.

Temperature (K)	293(2)
Empirical formula	$C_{17}H_{20}N_6O_6$
Formula weight	404.39
Wavelength (Å)	0.71073
Crystal system	triclinic
Space group	P-1
a (Å)	6.6889(2)
b (Å)	11.5839(3)
c (Å)	13.7723(3)
$\alpha(^{\circ})$	112.910(2)
$\beta(^{\circ})$	91.715(2)
$\gamma(^{\circ})$	104.582(2)
Volume (Å^3)	941.44(4)
Z	2
Calculated density (g/cm^3)	1.427
Absorption coefficient (mm^{-1})	0.111
F(000)	424
Crystal size (mm^3)	0.36 x 0.19 x 0.03
θ range for data collection (°)	1.62 - 28.38
Index ranges	$-8 \leq h \leq 8, -15 \leq k \leq 15, -18 \leq l \leq 18$
Reflections collected/unique	22267 / 4626
Completeness to θ=50°	99.9%
Transmission factors (min/max)	0.917/0.999
Data/restraints/parameters	4626/0/274
Goodness–of–fit on F^2	0.999
Final R indices [I> 2σ(I)]	0.0454 /0.1022
R indices (all data)	0.1039 /0.1266
Largest diff. peak and hole (e Å^{-3})	−0.249 /0.166
CCDC Number	CCDC 724893

The central guanidinium fragment of the cation is planar and the geometry is close to that expected for a central C_{sp2} atom. The bond lengths C1-N1 [1.336(2) Å] and C1-N2 [1.337(2) Å] are within the range expected for a delocalized C=N bond, while the C1-N3 bond length is slightly shorter [1.313(2) Å]. The phenyl rings are flat within 0.01 Å and have *anti-anti* conformation with respect to the unsubstituted N3 atom. The dihedral angle between the two phenyl rings is 59.87(10)° and the dihedral angles between the ring planes and the plane defined by the central guanidine fragment are 79.31(10)° (C2-C7) and 53.33(11)° (C8-C13).

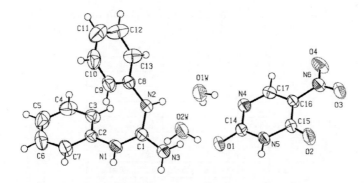

Figure 3. ORTEP diagram of diphenylguanidinium 5-nitrouracilate dihydrate. The thermal ellipsoids were drawn at the 50% probability level.

5-Nitrouracil is deprotonated at N1, thus negatively charged. The pyrimidine ring is almost planar, the weighted average absolute torsion angle is 1.3(1)° and the nitro group is rotated 6.04(19)° out of the ring plane. Higher distortion is seen in a similar compound, N, N', N''-Triphenylguanidinium 5-nitro-2,4-dioxo-1,2,3,4-tetrahydropyrimidin-1-ide, where the nitro group is rotated 11.4(2)° [15].

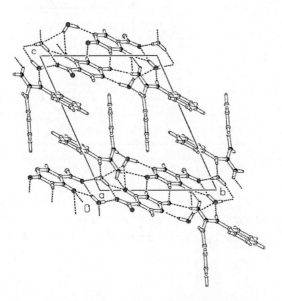

Figure 4. Packing of the molecules in the diphenylguanidinium 5-nitrouracilate dihydrate crystal structure. The dashed lines represent hydrogen bonds.

Table 5. Selected bond lengths (Å) and angles (°) for diphenylguanidinium 5-nitrouracilate dehydrate.

N1 – C2	1.416(2)
N2 – C8	1.417(2)
N4 – C17	1.320(2)
N4 – C14	1.364(2)
N5 – C14	1.365(2)
N5 – C15	1.385(2)
C1 – N1 – C2	128.47(14)
C1 – N2 – C8	127.86(14)
N3 – C1 – N1	118.86(15)
N3 – C1 – N2	117.82(15)
N1 – C1 – N2	123.32(16)
O4 – N6 – O3	121.28(14)

Table 6. H-bonding geometry for diphenylguanidinium 5-nitrouracilate dehydrate (Å, °).

Donor-H....Acceptor	D - H	H...A	D...A	D - H...A
N1-H1...O2W[i]	0.86	2.00	2.853(2	170
O1W-H1A ... N4	0.92(2	1.88(2	2.792(2	171(2)
O1W-H1B ... O2W	0.79(3)	2.49(3)	3.177(3)	145(3)
O1W-H1B ... O2[ii]	0.79(3)	2.50(3)	3.192(2)	147(3)
N2-H2 ... .O1W	0.86	1.90	2.752(2)	171
O2W-H2A ... O3[iii]	0.85(3)	2.10(3)	2.951(3)	176(3)
O2W-H2B ... O2[ii]	0.91(3)	1.98(3)	2.746(3)	140(2)
O2W-H2B ... O3 [ii]	0.91(3)	2.26(3)	3.000(2)	138(2)
N3-H3A ... O2[iv]	0.86	2.17	2.890(2)	141
N3-H3B ... O1	0.86	2.18	2.924(2)	144
N5-H51 ... O1[iv]	0.86	2.11	2.9426(19)	163

i: 1+x,y,z, *ii*: 1-x,1-y,-z , *iii*: x,-1+y,z, *iv*: 2-x,1-y,-z.

Assisted by the two water solvate molecules, anions and cations aggregate in slabs stacked along *c*. Dpg+ exhausts its capacity as a donor and the anion is very much involved in a donor/acceptor role, only O4 is left unperturbed (Figure. 4).

DISCUSSION

The dipole, the octupolar moment and the polarizability of protonated dpg molecules, and therefore the optical and dielectric properties of dpg salts, depend on the orientation of the rings, which justifies the need to determine accurate structural data for these compounds. At the present time sufficient structures have been reported so that common conformations or patterns can be identified. Table 7 contains a brief identification of all the structures analyzed in this review.

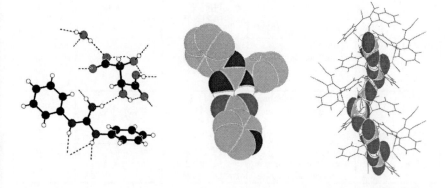

Figure 5. Left: Diphenylguanidinium tartrate hydrate (tartaric acid contains two chiral centres). Middle: Propeller-shape of Diphenylguanidinium nicotinate (nicotinic acid is bonded to the guanidinium fragment via H-bonds). Blue spheres represent nitrogen atoms, Red spheres represent oxygen atoms and black/grey spheres represent carbon atoms. Right: Helical arrangement of phosphite H-bonded anions in Diphenylguanidinium dihydrogenphosphite.

Table 7. Summary of crystallographic data for dpg and dpg$^+$ compounds.

CSD code	Coumpound name	Unit cell (Å, °)	Space group	Reference
CEYCIS	N,N'-Diphenylguanidinium formate	$a = 6.363(4)$ $b = 12.535(4)$ $c = 16.641(3)$	$P\,2_1cn$ non-centrosymmetric	[16]
DPGGUAN02	N,N'-Diphenylguanidine	$a = 8.906(2)$ $b = 12.342(1)$ $c = 21.335(2)$ $\beta = 96.66(1)$	$P2_1/c$	[7]
DPGGUAN03	N,N'-Diphenylguanidine	$a = 9.003(5)$ $b = 12.699(3)$ $c = 20.522(8)$	$P2_12_12_1$ non-centrosymmetric chiral	[8]

Table 7. (Continued)

CSD code	Coumpound name	Unit cell (Å, °)	Space group	Reference
FUQDAW	bis(N,N'-Diphenylguanidinium) bis(hydrogen phosphonate) phosphonic acid monohydrate	$a = 10.961(3)$ $b = 12.204(4)$ $c = 14.128(4)$ $\alpha = 76.40(3)$ $\beta = 73.33(3)$ $\gamma = 64.50(4)$	$P\text{-}1$	[17]
GOLTOQ	N,N'-Diphenylguanidinium dihydrogen phosphate	$a = 22.8032(10)$ $b = 7.639(2)$ $c = 16.986(2)$ $\beta = 106.29(1)$	$C2/c$	[18]
HOFDAH	N,N'-Diphenylguanidinium (+)-L-hydrogen tartrate monohydrate	$a = 7.066(3)$ $b = 14.723(8)$ $c = 18.219(6)$	$P2_12_12_1$ non-centrosymmetric chiral	[19]
HOFDEL	bis(N,N'-Diphenylguanidinium) oxalate	$a = 19.784(9)$ $b = 11.101(9)$ $c = 12.384(4)$ $\beta = 106.59(2)$	$C2/c$	[20]
IBOWOL	Diphenylguanidinium hydrogen bifluoride	$a = 6.911(2)$ $b = 17.044(5)$ $c = 11.553(2)$ $\beta = 106.08(2)$	$P2_1/c$	[21]
IMIWAC	bis(N,N'-Diphenylguanidinium) hexachloro-tellurate	$a = 19.707(3)$ $b = 12.791(2)$ $c = 12.979(2)$	$Pna2_1$ non-centrosymmetric	[22]
IMIWEG01	bis(N,N'-Diphenylguanidinium) hexabromo-tellurate(iv)	$a = 19.670(2)$ $b = 13.069(2)$ $c = 13.272(2)$	$Pna2_1$ non-centrosymmetric	[23]
ISABAF	tris(syn-N,N'-diphenylguanidinium) hexafluoroferrate(III) hydrate	$a = 9.7211(11)$ $b = 9.7211(11)$ $c = 77.872(18)$ $\gamma = 120$	$R\text{-}3c$	[24]
MIVDIE	N,N'-Diphenylguanidinium dihydrogenphosphite	$a = 10.775(4)$ $b = 10.775(4)$ $c = 12.572(2)$	$P4_3$ non-centrosymmetric chiral	[25]
NEMNEY	bis(Diphenylguanidinium) pentafluorovanadate	$a = 21.036(4)$ $b=10.7044(13)$ $c = 12.481(2)$ $\beta=103.858(16)$	$C2/c$	[26]
NERVIQ	N,N'-Diphenylguanidinium 8-hydroxyquinoline-5-sulfonate	$a = 8.797(2)$ $b = 16.880(2)$ $c = 28.145(4)$	$Pbca$	[27]
NOZWEE	N,N'-Diphenylguanidinium nitrate	$a = 17.020(4)$ $b = 13.906(3)$ $c = 5.811(1)$	$Pna2_1$ non-centrosymmetric	[28]

Table 7. (Continued)

CSD code	Coumpound name	Unit cell (Å, °)	Space group	Reference
NUZSAC	N,N'-Diphenylguanidinium perchlorate	$a = 11.640(2)$ $b = 14.763(4)$ $c = 8.6175(17)$	$Pna2_1$ non-centrosymmetric	[29]
POVDUZ	N,N'-Diphenylguanidinium trifluoroacetate	$a = 10.153(2)$ $b = 13.824(3)$ $c = 11.673(3)$ $\beta = 110.14(2)$	$P2_1/c$	[30]
RESNEI	N,N'-Diphenylguanidinium hydrogenselenite monohydrate	$a = 6.360(1)$ $b = 19.272(2)$ $c = 12.604(2)$ $\beta = 93.75(1)$	$P2_1/n$	[31]
SANQUU	Diphenylguanidinium hydrogen oxalate	$a = 15.315(7)$ $b = 5.573(3)$ $c = 17.487(8)$ $\beta = 99.863(9)$	$P2_1/c$	[32]
SEHDEP	N,N'-Diphenylguanidinium phthalate	$a = 10.7714(12)$ $b = 8.440(2)$ $c = 11.420(3)$ $\beta = 112.172(12)$	$P2_1$ non-centrosymmetric chiral	[33]
SOXROM	N,N'-Diphenylguanidinium m-chlorobenzeneseleninate	$a = 7.277(1)$ $b = 12.793(2)$ $c = 20.013(4)$ $\beta = 94.733(12)$	$P2_1/c$	[34]
XAGWUX	bis(N,N'-Diphenylguanidinium) sulfate monohydrate	$a = 8.4603(4)$ $b = 8.8687(4)$ $c = 18.529(6)$ $\alpha = 95.99(3)$ $\beta = 91.78(3)$ $\gamma = 104.80(4)$	P-1	[35]
XANDAR	N,N'-Diphenylguanidinium acetate	$a = 11.886(4)$ $b = 10.672(4)$ $c = 12.198(9)$ $\beta = 106.06(4)$	$P2_1/a$	[36]
	N, N'-Diphenylguanidinium nicotinate hydrate	$a = 10.1356(2)$ $b = 11.6112(2)$ $c = 15.9465(3)$	$P2_12_12_1$ non-centrosymmetric chiral	
	N,N'-Diphenylguanidinium 5-nitrouracilate dihydrate	$a = 6.6889(2)$ $b = 11.5839(3)$ $c = 13.7723(3)$ $\alpha = 112.910(2)$ $\beta = 91.715(2)$ $\gamma = 104.582(2)$	P-1	

In the 25 analyzed dpg/dpg+ compounds 10 crystallize in non-centrosymmetric space groups (5 are chiral). In HOFDAH the chiralization is forced by introducing a chiral negative ion in the structure (L-tartrate, Figure. 5). In other cases, chiralization is spontaneous and it can be achieved either by the adoption of a 3-blade propeller like shape of dpg⁺ (the third blade comes from the negative ion strongly hydrogen bonded) like in N, N'-Diphenylguanidinium nicotinate hydrate (Figure. 5). Another possibility is the formation of helical chains, assembled via H-bonds, with a defined helicity, which is the case of MIVDIE where the phosphite ions are joined in helical chains with the dpg+ following the same mould.

In a crystal engineering perspective, some non-centrosymmetric groups are preferable over others: it has been shown that crystal point groups 1, 2, *m*, *mm*2 correspond to the highest phase-matchable coefficient if the constituting molecules are one-dimensional charge transfer systems [37]. For dpg/dpg+ crystals the correlation between the crystal space group and the maximization of the non-linear optical properties is not so straightforward since dpg/dpg+ is a three-dimensional charge transfer system.

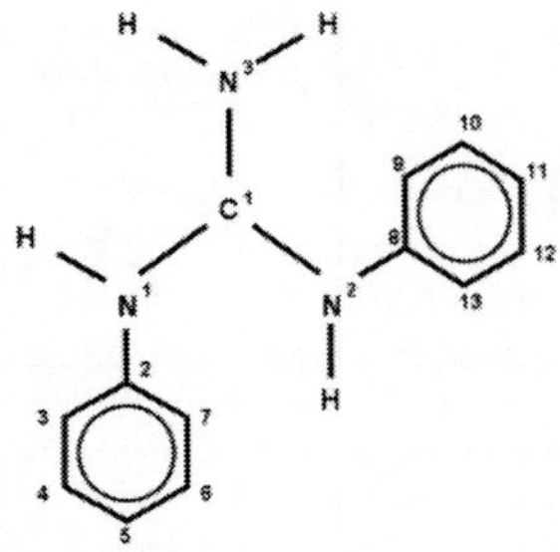

Scheme 4. Labelling scheme for dpg+ cation.

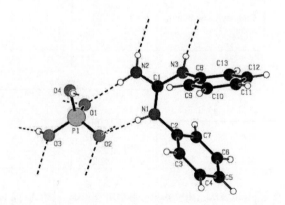

Figure 6. Molecular diagram of N,N'-Diphenylguanidinium dihydrogen phosphate, GOLTOQ.

Table 8. Ring conformation and related torsion angles.

CSD code	Ring Conformation		φ₁(°)		φ₂(°)		χ₁(°)		χ₂(°)	
CEYCIS	Syn–Syn		10.1(4)		14.3(4)		32.7(4)		36.7(4)	
DPGGUAN02	Syn–Anti	Syn–Anti	12.9(4)	-7.1(3)	-156.1(3)	165.5(2)	-89.4(4)	69.2(4)	-19.4(4)	1.4(5)
DPGGUAN03	Syn–Anti	Syn–Anti	-2.3(4)	10.7(4)	173.7(3)	-157.7(3)	-89.4(4)	69.2(4)	-19.4(4)	1.4(5)
FUQDAW	Syn–Anti	Syn–Syn	-15.9(6)	-8.9(7)	158.7(4)	-14.0(7)	-48.8(5)	56.0(6)	-43.9(6)	-45.6(7)
GOLTOQ	Anti–Anti		151.67(17		144.92(17)		-46.9(2)		-33.7(3)	
HOFDAH	Syn–Syn		-7.9(3)		-8.0(4)		-53.1(3)		-40.8(4)	
HOFDEL	Anti–Anti		151.18(17)		149.85(16)		-29.1(3)		-38.1(2)	
IBOWOL	Syn–Anti		-26.8(4)		166.6(3)		-38.2(4)		-37.0(4)	
IMIWAC	Syn–Syn		6.2(13)		11.4(10)		47.8(13)		61.9(10)	
IMIWEG01	Syn–Syn	Syn–Syn	7.7(13)	7.7(12)	6.0(15)	-9.2(11)	68.4(10)	-66.1(12)	55.0(13)	-46.6(14)
ISABAF	Syn–Syn		6.7(2)				39.8(3)			
MIVDIE	Syn–Anti		0.1(3)		170.41(17)		-87.8(2)		-43.3(3)	
NEMNEY	Anti–Anti		-152.3(4)		-150.3(4)		31.8(6)		35.9(6)	
NERVIQ	Syn–Syn		7.9(5)		0.6(6)		50.2(4)		-57.5(5)	
NOZWEE	Syn–Syn		-8.7(5)		-18.0(4)		-30.7(4)		-34.2(4)	
NUZSAC	Syn–Anti		-2.0(6)		174.7(4)		-68.9(5)		-80.1(5)	
POVDUZ	Syn–Syn		5.0(4)		9.4(4)		61.4(3)		48.6(4)	
RESNEI	Syn–Syn		-12.7(5)		-5.2(5)		-42.1(5)		-43.9(5)	
SANQUU	Syn–Syn		8.6(6)		18.3(6)		-53.4(5)		29.3(6)	
SEHDEP	Syn–Anti		6.7(4)		-163.0(2)		-65.3(4)		58.5(4)	
SOXROM	Syn–Anti		9.0(13)		-146.1(8)		38.2(13)		32.7(11)	

Table 8. (Continued)

CSD code	Ring Conformation		φ$_1$(°)		φ$_2$(°)		χ$_1$(°)		χ$_2$(°)	
	Syn-Syn	Anti-Anti								
XAGWUX	Syn-Syn		2.8(8)	152.7(5)	-1.5(10)	146.2(6)	-43.2(9)	-37.9(18)	37.3(9)	-27.6(9)
XANDAR		Anti-Anti	157.25(17)		152.44(17)		-37.6(3)		-33.4(3)	
Diphenylguanidinium nicotinate hydrate	Anti-Syn		158.96(18)		-13.0(3)		-53.0(3)		-47.4(3)	
Diphenylguanidinium 5-nitrouracilate dihydrate		Anti-Anti	-153.71(18)		-147.03(17)		19.2(3)		26.4(3)	

Table 8 gathers information about the torsion angles that define the dpg+ conformation changing significantly from cation to cation. The torsion angles are depicted in Scheme 2. The labelling scheme is shown in Scheme 4. N1 is chosen so that N1-C is smaller than N2-C. N3 is the unsubstituted nitrogen atom. C3 is chosen so that C3...N3 distance is smaller than C7...N3, C9 is chosen so that C9...N3 is smaller than C13-N3. In this way, φ_1 is defined as the N3-C1-N1-C2 torsion angle, φ_2 is N3-C1-N2-C8, χ_1 is C1-N2-C2-C3 and χ_2 is C1-N2-C8-C9. Most of φ angles lie in the -30 to $30°$ region, corresponding to a synperiplanar conformation or in the -180 to $-150°$, $+150$ to $+180°$ corresponding to the antiperiplanar conformation. GOLTOQ (Figure. 6) is one of the few exceptions with a φ_2 angle of $144.92(17)°$ corresponding to a $-$anticlinal conformation.

Figure 7 shows a plot of χ as a function of φ. A rough correlation can be seen: if φ is positive then χ is positive and if φ is negative then χ is negative. The crystal structure of XAGWUX contains two independent molecules with opposite conformations. One of the molecules is *syn-syn* the other is *anti-anti*. This occurrence frustrates any attempts to correlate the conformation of the cation with the acidic strength of the anion or other counter ion effects.

Table 9 collects information about the hydrogen bond network that is established in the compounds under discussion. With exception of the two diphenylguanidine polymorphs, orthorhombic and monoclinic, where there is a lack of acceptors, all the compounds show a full exhaustion of the guanidinium capacity as a proton donor. The most common situation is the H-bonding of molecules in slabs or layers, a situation that is easily understood by the planarity of the donor group and of the bulkiness of the phenyl rings that accommodate between layers. From all the patterns drawn by the H-bond network the most striking is that of ISABAF (Figure. 9). In this series of compounds H-bonds avoid making only discrete patterns but instead congregate to form chains or rings, thus better supporting the cohesion between molecules. A search for repeated synthons show a tendency for four dpg+ ions to aggregate around two small carboxylate ions like oxalate (HOFDEL), trifluoroacetate (POVDUZ) or acetate (XANDAR), Figure. 8. However diversity amongst the patterns is what stands out.

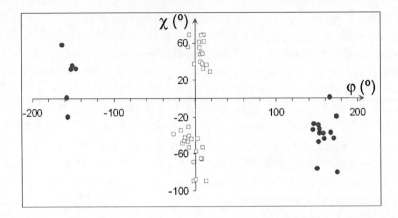

Figure 7. Plot of χ in function of φ, the circles correspond to rings oriented *anti* to the guanidinium moiety and squares correspond to rings oriented *syn*.

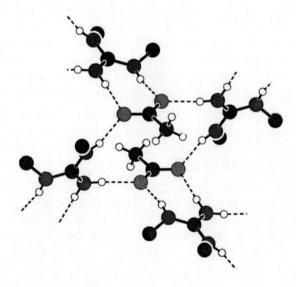

Figure 8. Partial packing diagram of diphenylguanidinium acetate, XANDAR. H-bonds are drawn as dashed lines, some atoms were omitted for clarity.

For the two polymorphs of the neutral diphenylguanidine the hydrogen bonding is markedly different. In the orthorhombic phase (DPGUAN03), the two symmetry independent molecules are linked by hydrogen bonds in infinite chains running along [100]. The hydrogen-bonding functionality of the two symmetry independent molecules is similar (Figure.10). The imino N atom of one molecule accepts two protons from the other, one donated by the NH_2 group and the other by the NH group. In the monoclinic phase (DPGUAN02), the imino N atom of each molecule accepts a single proton, donated either by the NH_2 or NH group of the other symmetry independent molecule. Therefore, and in contrast to the orthorhombic phase, the NH group acts as a donor in one molecule and as an acceptor in the other (Figure.10). Another difference in the hydrogen bonding of the two polymorphs concerns the role of the NH_2 groups. In the monoclinic phase, one molecule donates its protons and the other is only involved in very weak hydrogen bonds, the shortest N...N distance being 3.520(2) Å. Interestingly, although the hydrogen bonding network is more extensive in the orthorhombic than in the monoclinic crystals, the full potential for hydrogen bonding of the dpg molecules is not fulfilled in either polymorph.

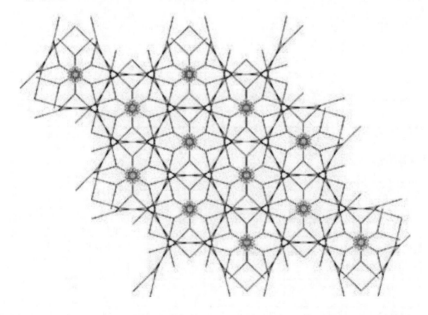

Figure 9. Packing diagram of tris(N,N'-diphenylguanidinium) hexafluoroferrate(III) hydrate, ISABAF, viewed along c. To maximize visual impact, H-bonds and covalent bonds were drawn the same way.

Table 9. H-Bonds motifs, (if the full capacity of the guanidinium fragment as a donor is being used, the second column hits yes).

CSD code	Fully exhausted	Dimensionality	Brief Description	Motifs
CEYCIS	yes	2D	slabs parallel to ab plane	small and large rings if viewed along the c axis
DPGGUAN02	no	1D	chains along a	8 element repetition
DPGGUAN03	no	1D	chains along a	8 element repetition
FUQDAW	yes	3D	adjacent tubes along a	large rings if viewed along a
GOLTOQ	yes	2D	slabs parallel to bc	chains of phosphate ions running along the c axis, C(5)
HOFDAH	yes	3D		small and large rings if viewed along a
HOFDEL	yes	2D	slabs parallel to ab	4 DPG H-bonded to one oxalate ion
IBOWOL	yes	2D	layers parallel to ac	small and large rings if viewed along b
ISABAF	yes	3D		flower-like motif if viewed along c
MIVDIE	yes	3D		helical chains of phosphite ions alonc c axis, C(5)
NEMNEY	yes	2D	layers parallel to bc	4 DPG H-bonded to one vanadate ion
NERVIQ	yes	2D	layers parallel to ab	chains of sulfonate ions along b axis, C(8)
NOZWEE	yes	2D	layers parallel to bc	small and large rings if viewed along b
NUZSAC	yes	3D		3 types of rings if viewed along c
POVDUZ	yes	2D	layers parallel to bc	4 DPG differently H-bonded to two negative ions
RESNEI	yes	3D		large rings if viewed along a
SANQUU	yes	2D	layers parallel to bc	chains of oxalate ions along b axis, C(5)
SEHDEP	yes	2D	layers parallel to bc	3 types of rings if viewed along a
SOXROM	yes	1D	chains along a	7 element repetition
XAGWUX	yes	2D	slabs parallel to ab	water and sulfate ions aggregate in dimers
XANDAR	yes	2D	layers parallel to ab	4 DPG H-bonded to two negative ions

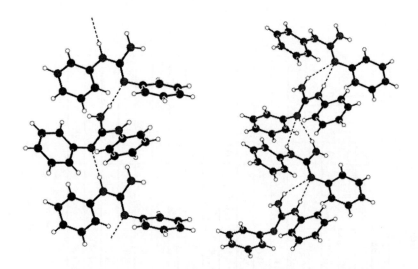

Figure 10. Chain formation in the monoclinic phase of diphenylguanidine (left) and in the orthorhombic phase of diphenylguanidine (right).

The C-N bond lengths lie between the values for single and double C-N distances. This partial double bond character is consistent with the new type of aromaticity called Y-aromaticity often attributed to the guanidinium ion. Protonation of guanidine leads to a highly symmetric ion, with six π-electrons, for which three equivalent resonance structures are possible. The resulting delocalization is responsible for the fact that it is energetically very favorable for guanidine to accept a proton, making it a strong organic base [38]. Table 10 also shows a calculated HOMA (harmonic oscillator measurement of aromaticity) index defined so that HOMA=0 for a non-aromatic system and HOMA=1 when full delocalization of π-electrons occur [38]. Inspection of Table 10 shows the expected aromaticity increase upon protonation of the guanidine fragment.

In a study of tautomeric heterocycle systems Zubatyuk et al. [39] concluded that the position of tautomeric equilibrium is controlled by the aromaticity of the heterocycle. They have also shown a strong correlation between the strength of the a intramolecular hydrogen bond and the aromaticity of a resonant spacer. In DPGUAN03, where two symmetry independent molecules, establish similar hydrogen bonds, the lowest aromaticity corresponds to the strongest intermolecular H-bonds.

Table 10. C-N bond lengths and HOMA index.

CSD code	C1-N1(Å)		C1-N2(Å)		C1-N3(Å)		HOMA index	
CEYCIS	1.335(3)		1.337(3)		1.323(3)		0.996	
DPGGUAN02	1.278(4)	1.287(4)	1.386(3)	1.374(4)	1.358(3)	1.358(4)	0.831	0.835
DPGGUAN03	1.292(4)	1.278(3)	1.367(3)	1.368(3)	1.357(4)	1.335(4)	0.856	0.910
FUQDAW	1.315(6)	1.333(5)	1.341(7)	1.337(6)	1.324(6)	1.317(6)	0.995	0.980
GOLTOQ	1.342(2)		1.343(2)		1.311(2)		0.979	
HOFDAH	1.330(3)		1.332(3)		1.314(3)		0.987	
HOFDEL	1.333(2)		1.341(2)		1.314(2)		0.986	
IBOWOL	1.336(3)		1.340(3)		1.311(3)		0.983	
IMIWAC	1.322(8)	1.313(11)	1.332(9)	1.327(10)	1.328(10)	1.307(10)	0.986	0.971
IMIWEG01	1.329(9)	1.307(8)	1.330(8)	1.335(8)	1.317(8)	1.304(8)	0.969	0.972
ISABAF	1.335(2)				1.316(3)		0.990	
MIVDIE	1.327(3)		1.336(3)		1.324(3)		0.995	
NEMNEY	1.333(4)		1.340(5)		1.315(5)		0.987	
NERVIQ	1.324(5)		1.334(5)		1.317(5)		0.988	
NOZWEE	1.328(3)		1.348(3)		1.310(3)		0.974	
NUZSAC	1.322(5)		1.341(5)		1.322(5)		0.989	
POVDUZ	1.329(3)		1.334(3)		1.318(3)		0.991	
RESNEI	1.336(2)		1.342(2)		1.317(2)		0.988	

Table 10. (Continued)

CSD code	C1-N1(Å)		C1-N2(Å)		C1-N3(Å)		HOMA index	
SANQUU	1.324(4)		1.339(4)		1.321(4)		0.990	
SEHDEP	1.327(3)		1.337(3)		1.313(3)		0.984	
SOXROM	1.332(10)		1.351(10)		1.325(10)		0.988	
XAGWUX	1.325(8)	1.323(7)	1.344(7)	1.357(7)	1.300(7)	1.309(8)	0.956	0.961
XANDAR	1.333(2)		1.353(2)		1.305(2)		0.961	
DPG nicotinate hydrate	1.338(2)		1.338(2)		1.315(2)		0.987	
DPG 5-nitrouracilate dihydrate	1.336(2)		1.337(2)		1.313(2)		0.986	

However in the dpg/dpg+ systems, the influence of hydrogen bonds between the guanidine group and other molecules is not the sole factor influencing aromaticity, since the interaction between the guanidinium delocalized π-orbitals and the ring π-orbitals is also very important.

In order for a material to have useful and highly efficient nonlinear optical properties, the constituting molecules need to exhibit large second-order molecular hyperpolarizabilities. Furthermore, the molecules need to align non-centrosymmetrically in a favorable orientation such that the molecular contribution can be maximized in the bulk. Calculations using the PM6 Hamiltonian [40] available in MOPAC2009 [41] yields small dipolar moments and similar hyperpolarizabilities for all the dpg/dpg+ molecules. The *syn-syn* conformation results in smaller than 1 Debye dipolar moments, the *syn-anti* outcome is less than 5 D and *anti-anti* conformation rises up to ~9 D. All the dpg/dpg+ molecules have higher octupolar character, a property that can be accessed by calculating the ρ^{3D} parameter as defined in reference 42. The higher ρ^{3D}, the more octupolar is the character of the molecule, GOLTOQ has the higher ρ^{3D} (~10), an interesting feature since GOLTOQ is one of the few compounds reviewed that has an anticlinal conformation.

CONCLUSIONS

The analysis of these 25 compounds confirms the flexibility of dpg/dpg+. It can adopt a propeller like conformation and induce the formation of chiral crystals. It establishes predominantly two-dimensional H-bonding networks with the counter ion. From the perspective of the crystal engineer, an anticlinal conformation would maximize the non-linear properties of the molecule. However the non-linear optical response depends greatly on the molecular alignment. Non-centrosymmetric dispositions have not yet been achieved for dpg with an *anti-anti* conformation. The best strategy to attain NLO samples is by inducing a *syn-anti* conformation where a propeller structure might induce a chiral disposition of molecules and thus a non-centrosymmetric order. Another successful approach is to force crystallization with a chiral counter-ion.

REFERENCES

[1] Olney, JW; Labruyere, J; Price, MT. *Science*, 1989, 244, 1360-1362.

[2] Weber, E; Sonders, M; Quarum, M; McLean, S; Pou, S; Keane, JFW. Proc. Natl. *Acad. Sci. USA*, 1986, 83, 8784-8788.

[3] Largent, BL; Wikstrom, H; Gundlach, AL; Snyder, SH. *Mol. Pharmacol*, 1987, 32, 772-784.

[4] Zyss, J; Pecaut, J; Levy, JP; Masse, R. *Acta Crystallogr*, 1993, B49, 334-342.

[5] Kanamori, K; Roberts, JD; J. Am. *Chem. Soc.*, 1983, 105, 4698-4701.

[6] Botto, RE; Schwartzs, JH; Proc. Natl. *Acad. Sci. USA*, 1980, 77, 23-25.

[7] Tanatani, A; Yamaguchi, K; Azumaya, I; Fukutomi, R; Shudo, K; Kageshika, H; *J. Am. Chem. Soc.*, 1998, 120, 6433-

[8] Paixão, JA; Beja, AM; Silva, PSP; Ramos Silva, M; da Veiga, LA. *Acta Cryst. C*, 1999, C55, 1037-

[9] Alagona, G; Ghio, C; Nagy, P; Durant, GJ. *J. Phys. Chem.*, 1995, 98, 5422-5430.

[10] Nagy, P; Durant, GJ; *J. Chem. Phys.*, 1996, 104, 1452-1463.

[11] Allen, FH; *Acta Cryst. B*, 2002, B58, 380-388.

[12] Sheldrick, GM; *Acta Cryst. A*, 2008, A64, 112-122.

[13] Allen, FH; Kennard, O; Watson, DG; Brammer, L; Orpen, AG; Taylor, R. *J. Chem. Soc. Perkin Trans*, 1987, 2 , S1-19.

[14] Puccetti, G; Perigaud, A; Badan, J; Ledoux, I; Zyss, J. *J. Opt. Soc. Am. B*, 1993, 10, 733-744.

[15] Pereira Silva, PS; Domingos, SR; Ramos Silva, M; Paixão, JA; Beja, AM. *Acta Cryst*, 2008, E64, o1082-o1083.

[16] Paixão, JA; Matos Beja, A; Ramos Silva, M; Alte da Veiga, L. *Z. Kristallogr. NCS*, 1999, 214, 475.

[17] Paixão, JA; Matos Beja, A; Ramos Silva, M; Alte da Veiga, L. *Z. Kristallogr. NCS*, 2000, 215, 352.

[18] Silva, PSP; Paixão, JA; Beja, AM; Ramos Silva, M; da Veiga, LA. *Acta Cryst. C*, 1999, C55, 1096-1099.

[19] Paixão, JA; Silva, PSP; Beja, AM; Ramos Silva, M; de Matos Gomes, E; Belsley, M. *Acta Cryst. C*, 1999, C55, 1287-1290.

[20] Paixão, JA; Beja, AM; Ramos Silva, M; da Veiga, LA. *Acta Cryst. C*, 1999, C55, 1290-1292.

[21] Ramos Silva, M; Paixão, JA; Beja, AM; da Veiga, LA. *J. Fluorine Chem.*, 2001, 107, 117-120.

[22] Mirochnik, AG; Bukvetskii, BV; Storozhuk, TV; Karasev, VE. Zh. Neorg. *Khim*, 2003, 48, 582-591.

[23] Bukvetskii, BV; Mirochnik, AG; Zh.Neorg. *Khim*, 2005, 46, 694-698.

[24] Ramos Silva, M; Beja, AM; Costa, BFO; Paixão, JA; da Veiga, LA. *J. Fluorine Chem.*, 2000, 106, 77-81.

[25] Paixão, JA; Matos Beja, A; Ramos Silva, M; Alte da Veiga, L. Z. Kristallogr. *NCS*, 2001, 216, 416.

[26] Ramos Silva, M; Matos Beja, A; Paixão, JA; Alte da Veiga, L. Z. Kristallogr. *NCS*, 2001, 216, 261.

[27] Silva, PSP; Ramos Silva, M; Beja, AM; Paixão, JA; *Acta Cryst. E*, 2006, E62, o5913.

[28] Paixão, JA; Pereira Silva, PS; Beja, AM; Ramos Silva, M; da Veiga, L. A; *Acta Cryst. C*, 1998, C54, 805-808.

[29] Paixão, JA; Pereira Silva, PS; Matos Beja, A; Ramos Silva, M; Alte da Veiga, L. Z. Kristallogr. *NCS*, 1998, 213, 419.

[30] Paixão, JA; Pereira Silva, PS; Beja, AM; Ramos Silva, M; da Veiga, L. A. *Acta Cryst. C*, 1998, C54, 1484-1486.

[31] Paixão, JA; Beja, AM; Ramos Silva, M; de Matos Gomes, E; Martín-Gil, J; Martín-Gil, FJ. *Acta Cryst. C*, 1997, C53, 1113-1115.

[32] Hu, ML; Wang, WD. *Acta Cryst. E*, 2006, E61, o1408.

[33] Silva, PSP; Ramos Silva, M; Beja, AM; Paixão, JA; *Acta Cryst. E*, 2006, E62, o1067.

[34] Antolini, L; Marchetti, A; Preti, C; Tagliazucchi, M; Tassi, G; Tosi, G; Aust. *J. Chem.*, 1991, 44, 1761-1769.

[35] Matos Beja, A; Paixão, JA; Ramos Silva, M; Alte da Veiga, L; de Matos Gomes, E; Martín-Gil, J. Z. Kristallogr. *NCS*, 1998, 213, 655.

[36] Matos Beja, A; Paixão, JA; Ramos Silva, M; Alte da Veiga, L. Z. Kristallogr. *NCS*, 2000, 215, 579.

[37] Zyss, J; Oudra, JL; *Phys. Rev. A*, 1982, 26, 2028-2048.

[38] Raczyńska, ED; Cyrański, MK; Gutowski, M; Rak, J; Gal, JF; Maria, PC; Parowska, M; Duczmal, K. *J. Phys. Org. Chem.*, 2003, 16, 91-106.

[39] Zubatyuk, R; Shishkin, OV; Gorb, L; Leszczynski, J. *J. Phys. Chem. A*, 2009, 113, 1943-2952.

[40] Stewart, JJP. *J. Mol. Mod*, 2007, 13, 1173-1213.

[41] MOPAC2009, James, JP. Stewart, Stewart Computational Chemistry, Version 9.045W web: HTTP://OpenMOPAC.net.

[42] Zyss, J. *J. Chem. Phys.*, 1993, 98, 6583-6599.

In: Chemical Crystallography
Editor: Bryan L. Connelly, pp. 131-151
ISBN: 978-1-60876-281-1
© 2010 Nova Science Publishers, Inc.

Chapter 4

CONSTRUCTION AND STRUCTURE OF METAL-ORGANIC FRAMEWORKS WITH SPECIFIC ION-EXCHANGE PROPERTY

*Man-Sheng Chen, Zhi Su, Shui-Sheng Chen and Wei-Yin Sun**

Coordination Chemistry Institute, State Key Laboratory of Coordination Chemistry, School of Chemistry and Chemical Engineering, Nanjing National Laboratory of Microstructures, Nanjing University, Nanjing 210093, China

ABSTRACT

Remarkable progress has been achieved in the area of metal-organic frameworks (MOFs) in recent years not only due to their diverse topology and intriguing structures but also owing to their interesting physical and chemical properties. MOFs with specific ion exchange property have attracted great attention for their potential application in molecular/ionic recognition and selective guest inclusion. Despite the difficulty in predicting the structure and property of MOFs, the increasing knowledge regarding the synthesis methods and characterization techniques has largely expanded for the rational designs.

---

* Corresponding address: E-mail sunwy@nju.edu.cn.

In this chapter the recent works in cation/anion exchange with zero- (0D), one- (1D), two- (2D) and three-dimensional (3D) frameworks from our and other groups will be highlighted. Cation exchange mainly concentrates on the metal ions and organic cations such as M^{m+}, $[M(H_2O)_n]^{m+}$, $[Me_2NH_2]^+$, etc., while anion exchange comprises the majority of the counteranions, e.g., ClO_4^-, NO_3^-, BF_4^-, and so on. The functions of the exchanged compounds, i.e., enhancement of gas adsorption and photoluminescence, were greatly reformed.

INTRODUCTION

Metal-organic frameworks (MOFs) with varied structures such as zero-dimensional (0D) cages, capsules, boxes or polyhedra, one-dimensional (1D) chains and tubes, two-dimensional (2D) layers and three-dimensional (3D) frameworks have been constructed in the past years and the results showed that the nature of organic linkers and the geometric requirement of metal centers are crucial in determining the structures and properties of MOFs [1-16]. Since Dunbar and her co-workers reported the complex $[Ni_4(\mathbf{1})_4(CH_3CN)_8][BF_4]_8 \cdot 4CH_3CN$ [$\mathbf{1}$ = 3,6-bis(2-pyridyl)-1,2,4,5-tetrazine, Figure 1] with BF_4^- as an anion template in 1999 [17], the anion effects as well as the specific ion exchange property of MOFs have attracted great attention from chemists and became an energetic field of research for their potential applications in molecular/ionic recognition, selective guest inclusion and so on [18-22]. There are two different kinds of ion exchange in MOFs with different framework structures, namely, cation and anion exchanges. The former mainly concentrates on the metal and organic cations such as M^{m+}, $[M(H_2O)_n]^{m+}$, $[Me_2NH_2]^+$, etc., while the latter comprises the majority of the counter-anions—e.g., ClO_4^-, NO_3^-, BF_4^-—and few coordinating anions—e.g., from SO_4^{2-} to Cl^- [23]. In addition, it is known that the framework structures of MOFs are greatly dependent on the organic linkers and the metal ions—for example, the flexibility of ligands, the arrangement of the binding sites in the ligand, as well as the geometric requirements of the metal ions. Four typical multi-imidazole and pyridine containing ligands used in the construction of recently reported MOFs are schematically shown in Figure 1.

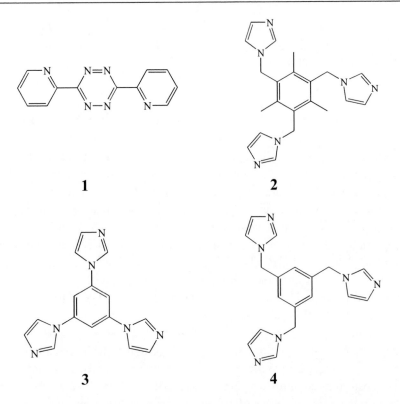

Figure 1. Schematic drawing of ligands with multi-pyridine or imidazole groups.

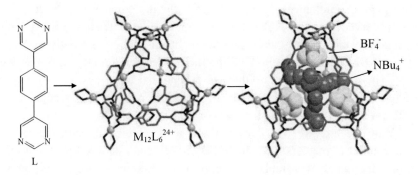

Figure 2. The formation and structure of the 0D cationic cage of **5** and its host-guest complex.

In this chapter, recent works with cation and anion exchange of 0D, 1D, 2D and 3D frameworks are introduced and, accordingly, the contents of this chapter are divided into four sections on the basis of the dimensionality of MOFs with ion exchange properties, and their relevant specific structures are included in each section.

ZERO-DIMENSIONAL CAGE-LIKE COMPOUNDS

One of the most challenging research fields in modern chemistry is the design and synthesis of multifunctional compounds and materials with predictable structures and properties. The cage-like compounds constructed from metal coordination with rigid or flexible organic ligands possess specific inner cavities that can have interesting properties such as molecular/ionic recognition, catalyzing reactions and ion exchange [24-31]. For instance, a complex $\{[Pd(en)]_{12}L_6\}(OTf)_{24}$ (5) (en = ethylenediamine, OTf = triflate), obtained by self-assembly reaction of the rigid ligand 1,4-bis(3,5-pyrimidyl)benzene (L) with cis-protected Pd(II) salt of $[Pd(en)](OTf)_2$, has 0D cationic $\{[Pd(en)]_{12}L_6\}^{24+}$ cage evidenced by the ^1H NMR spectral measurements as well as the CSI-MS (cold-spray ionization mass spectroscopy) in the solution [32]. It is interesting that the large inner cavity of the cage gives it a unique inclusion property: when NBu_4BF_4 was mixed with 5, the host-guest complex was formed with a triple layered onion-like structure in which the anionic sphere of BF_4^- is sandwiched between the cationic host cage and NBu_4^+, as illustrated in Figure 2. Furthermore, when the NEt_4^+ and NBu_4^+ were simultaneously mixed with 5 for the competition experiment, it was observed that the NEt_4^+ was first encapsulated arising from a kinetic selection of the smaller guest; then the included NEt_4^+ was slowly exchanged by the NBu_4^+ to exclusively yield the thermodynamic product after one week [32]. The results demonstrate that the exceptional host-guest behavior would lead to a new kind of pairwise selective sensing where a host recognizes two different species only when they coexist [32].

With regard to the construction of discrete polyhedra or infinite coordination frameworks, we and others have been interested in the reactions of organic ligands with imdazole groups and varied metal salts. The first Xray crystal structural that determined the M_3L_2 cage with a relatively large inside cavity was obtained by reaction of imidazole-containing ligand, 1, 3, 5-tris(imidazol-1-ylmethyl)benzene (4, Figure 1), with zinc(II) acetate dihydrate and was found to have guest inclusion property [33]. Furthermore, the interior cavity of the cationic cages has been certified to be an electrophilic microspace, which can include anionic or neutral species [34]. As a typical example, the anion exchange property of the M_3L_2 cage was well studied [35]. When the tripodal ligand 1,3,5-tris(imidazol-1-ylmethyl)-2,4,6-trimethyl-benzene (2, Figure 1) was reacted with silver(I) tetrafluoroborate salt in ethanol, the M_3L_2 type cage, $[BF_4{\subset}Ag_3(\mathbf{2})_2][BF_4]_2$ (6) was obtained [35]. The powdered sample of 6 was suspended in aqueous solution of excess $NaClO_4$ with stirring to allow anion exchange, and the results of spectroscopic measurements demonstrated that the BF_4^- was completely exchanged by ClO_4^-, which further demonstrated that the cationic cage can act as host for the specific anion and facilitate the anion exchange (Figure 3).

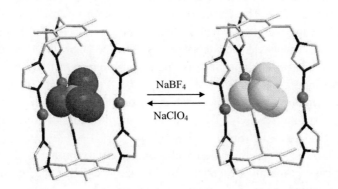

Figure 3. Schematic showing the anion exchange inside the cavity of the M_3L_2 cationic cage.

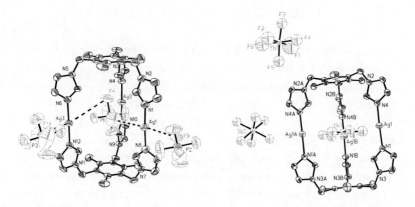

Figure 4. Crystal structures of the Ag_3L_2 cages $[PF_6{\subset}(Ag_3(\mathbf{2})_2](PF_6)_2$ (left) and $\{SbF_6{\subset}[Ag_3(\mathbf{2})_2](SbF_6)_2\}$ (right) with encapsulation of octahedral anion of PF_6^- or SbF_6^-.

To further investigate the anion effect on the formation of the cage as well as the size and shape in the anion inclusion and exchange, two M_3L_2 cage-like compounds $[PF_6 \subset (Ag_3 \ (2)_2] \ (PF_6)_2 \ (7)$ and $\{SbF_6 \subset [Ag_3(2)_2] \ (SbF_6)_2\}\cdot H_2O\cdot 1.5CH_3OH \ (8)$ were prepared and the results showed that the octahedral anions of PF_6^- and SbF_6^- can also be included in the inner cavity of the cage (Figure 4) which is the same as the tetrahedral ones of BF_4^-, ClO_4^- (Figure 3). It is interesting that, in the anion exchange reactions, the BF_4^- and ClO_4^- can be exchanged reversibly and the BF_4^- or ClO_4^- inside the cage can also be exchanged by PF_6^-, however, the PF_6^- inside the cage can not be exchanged by BF_4^- or ClO_4^-, only the ones outside the cage can be exchanged, while in the case of SbF_6^-, none of them (both inside and outside the cage) can be exchanged [36]. Therefore, the selective anion-inclusion order of the $[Ag_3(2)_2]^{3+}$ cage was found to be $SbF_6^- > PF_6^- > BF_4^-/ClO_4^-$. Similar study reported by Dalcanale and co-workers showed that the cavitand-based coordination cages obtained by assembly reactions of tetracyano cavitands with $[M(dppp)](OTf)_2$ [M = Pd(II) (9), Pt(II) (10), dppp = 1,3-bis(diphenylphosphino)propane] can encapsulate anions with different size and shape, such as $CF_3SO_3^-$, BF_4^- and PF_6^-, but the order of the encapsulation selectivity is $BF_4^- > CF_3SO_3^- >> PF_6^-$ [37].

It can be seen that the anion inclusion selectivity in the cavity of the 0D cationic cage is mainly dependent on the size and shape of the anion fitted with the size and shape of the cavity of the cage.

ONE-DIMENSIONAL CHAINS

Coordination polymers based on transition metal salts interacting with exo-polydentate N-donor ligands, for example 4,4'-bipyridine (bpy), have diverse structures and dimensionalities of 1D, 2D and 3D. It is noteworthy that in addition to the coordination interactions, coordination polymers with 1D or 2D structures can further linked together by hydrogen bonding and/or π-π interactions to generate the final 3D structure. The non-coordinated counter anions usually locate among the vacancy of the cationic polymers and only weakly interact with the cationic framework through the hydrogen bonding interactions, and as a result their removal or exchange does not disrupt the coordination interactions supporting the integrity of the polymers [38, 39]. In these cases, the anion can be considered as a guest species, similar to a solvent molecule trapped in the vacancy of the polymeric framework. For example, the reported 1D railroad-like coordination polymer $[Ni(bpy)_{2.5}(H_2O)_2]$ $(ClO_4)_2 \cdot 1.5bpy \cdot 2H_2O$ (**11**) has a 11 × 11 Å pore occupied perchlorate anions (Figure 5) which can be exchanged by PF_6^- ions confirmed by the IR spectral data: the disappearance of the characteristic bands of ClO_4^- (1165 - 1097 cm^{-1}) and the appearance characteristic ones of PF_6^- (850 and 569 cm^{-1}) [40]. And the crystalline of the compound remained upon the anion exchange as indicated by the sharp peaks observed in the Xray diffraction pattern of the exchanged sample.

Another good example is $[Zn_2(bpy)_3(H_2O)_8(ClO_4)_2(paba)_2] \cdot 2bpy \cdot 4H_2O$ (**12**) obtained by assembly reaction of *p*-aminobenzoate (paba) and bpy with $Zn(NO_3)_2 \cdot 6H_2O$ and $NaClO_4$ [41]. The 1D zigzag chain structure of **12** was formed by coordination and π-π interactions and the anion exchange of **12** was confirmed by single crystal Xray diffraction. When single crystals of **12** were immersed in an aqueous NH_4PF_6 solution over one week, single crystals of **12a** was obtained, and its structure was determined under the same initial conditions used for determining the structure of **12**. The results of crystallographic analysis (Figure 6) demonstrate that the perchlorate anions in compound **12** were completely replaced by slightly larger hexafluoro-phosphate anions, giving rise to an increase in the unit cell size of 113.6 $Å^3$ for **12a**. Moreover, single crystals of **12a** were immersed again in an aqueous $NaClO_4$ solution over one week to give single crystals of **12b**, which have the same structure as that of **12**. Meanwhile, the IR data of sample **12a** also show the disappearance of intense ClO_4^- bands and the appearance of intense PF_6^- ones indicating that the sample has undergone complete anion exchange. The data demonstrate that a completely reversible anion exchange between the ClO_4^- and PF_6^- can be achieved for complex **12**.

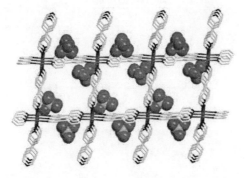

Figure 5. The packing structure of **11** with anions showing in space-filling mode; the coordinated water and non-coordinated bpy molecules were omitted for clarity.

In addition to the anion exchange of the 1D chain polymers with transition metal centers as mentioned above, a recently reported lanthanide based 1D coordination polymers, $\{[Ln_2(C_{16}H_{10}O_4N_4)_4Mn(H_2O)_6]\cdot x(H_2O)\}_n$ [$C_{16}H_{10}O_4N_4$ = 3, 3'-(4-amino-4H-1, 2, 4-triazole-3, 5-diyl)dibenzoate, Ln(III): La **(13)**, Gd **(14)**, Eu **(15)**, and Tb **(16)**], consisting of the nanosized Ln_2L_4 cage-like sub-units which encapsulate $[Mn(H_2O)_6]^{2+}$ cations [42]. It is interesting that such 1D polymers show the cation exchange property as illustrated in Figure 7. The investigations revealed that these 1D coordination polymers remain intact upon reversible cation exchange with different cations, which was characterized by the ion-dependent photoinduced emission spectra and Xray powder diffraction patterns. The results provide a promising and practical approach to access the multifunctional luminescent materials that can provide tunable emissions by controlling the type of guest species.

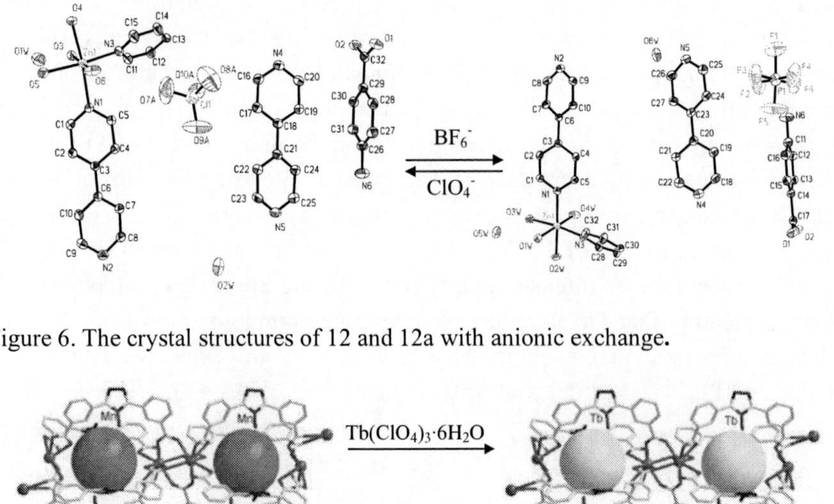

Figure 6. The crystal structures of 12 and 12a with anionic exchange.

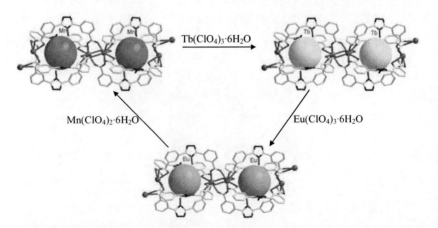

Figure 7. Cation exchange of 1D coordination polymers.

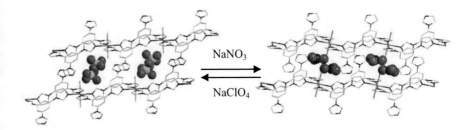

Figure 8. Reversible anion exchange between the complexes **17** and **18**.

TWO-DIMENSIONAL NETWORKS

Metal-ligand coordination based frameworks have been given particular attention due to their potential ability in selective inclusion and transportation of ions and/or molecules as well as in the catalysis of specific chemical reactions [43-45]. However, the control of the dimensionality and structure of the metal-organic coordination polymers is still a challenge since the assembly reactions can be influenced by several factors and the ancillary ligation of solvent or anion may result in low dimensionality even with the multi-functional ligands [46]. In order to investigate the influence of organic ligands on the structure and ion exchange property of the complexes, we designed and synthesized rigid and flexible tripodal ligands: 1,3,5-tris(1-imidazolyl)benzene (**3**) and 1,3,5-tris(imidazol-1-ylmethyl)benzene (**4**, Figure 1). A series of coordination polymers with 2D layered structure and anion exchange property were obtained [47-50]. For example, as evidenced by IR spectroscopy, elemental analysis and Xray powder diffraction, reversible anion exchanges were observed between the complexes $[Mn(3)_2(H_2O)_2](ClO_4)_2 \cdot 2H_2O$ (**17**) and $[Mn(3)_2(H_2O)_2](NO_3)_2$ (**18**) (Figure 8) without destruction of the frameworks, which implied that the 2D networks **17** and **18** can act as cationic layered materials for the anion exchange [47]. The other 2D MOFs, e.g., $[Ag(2)]ClO_4$, (**19**), $[Mn(2)_2]SO_4 \cdot 16H_2O$ (**20**) and $[Zn(2)_2](NO_3)_2 \cdot 2MeOH$ (**21**) also showed anion exchange properties [48-50].

The flexible ligand **4** can adopt different conformations when it reacts with metal salts [51]. For instance, ligand **4** reacted with varied zinc(II) salts ZnX [X = (BF_4)_2, SO_4, Cl_2, Br_2, I_2] to afford a series of coordination polymers with different structures. Coordination polymer $\{[Zn(4)_2](BF_4)_2\}_n$ (**22**) has an infinite 2D cationic double layered structure in which the ligand **4** has *cis, cis, cis*-conformation, and the tetrafluoroborate anions are located in the voids of two adjacent cationic layers. The results of elemental analysis and IR spectral measurements confirmed that the BF_4^- anions in **22** can be completely exchanged by NO_3^- or NO_2^- anions to give compounds **22A** and **22B**, respectively, as shown Figure 9.

Suh et al. reported Ag(I)-polynitrile networks for anion exchange, [Ag(EDTPN)](CF$_3$SO$_3$) (**23**) and [Ag(EDTPN)](ClO$_4$) (**24**) with different 2D layer, and [Ag(EDTPN)(NO$_3$)] (**25**) with 1D chain structures were obtained by assembly reactions of ethylenediaminetetrapropionitrile (EDTPN) with the corresponding silver(I) salts [46]. Complex **23** was immersed in the aqueous solution of NaNO$_3$ and NaClO$_4$, respectively, the structure of **23** was converted to that of **24** and **25** through the anion exchange of $CF_3SO_3^-$ to NO_3^- and ClO_4^-, respectively. Complex **25** can also be transformed to **23** and **24** by immersion in the LiCF$_3$SO$_3$ and NaClO$_4$ aqueous solution accompanying with the NO_3^- exchanged by $CF_3SO_3^-$ and ClO_4^-, respectively, however, the ClO_4^- anion in **24** can not be exchanged by $CF_3SO_3^-$ or ClO_4^- to generate **23** and **25** (Figure 10). In addition, when 23 was immersed in the solution containing both NO_3^- and ClO_4^- for the anion exchange competition, the results showed that the $CF_3SO_3^-$ in **23** was only exchanged by ClO_4^-, indicating that the ClO_4^- induced network 24 is thermodynamically more stable than the NO_3^- induced 1D chain structure of 25. The results showed the anion-induced formation and further transformation of the supramolecular structure. Furthermore, the anions $CF_3SO_3^-$ and ClO_4^- in 23 and 24 locate in the vacancy of the 2D layers, while in the case of **25** the NO_3^- anion coordinated with the metal atom. Namely, the study provided nice example that the coordinated NO_3^- anions in **25** can be exchanged by $CF_3SO_3^-$ and ClO_4^-, while the non-coordinated ClO_4^- in **24** can not be exchanged [46].

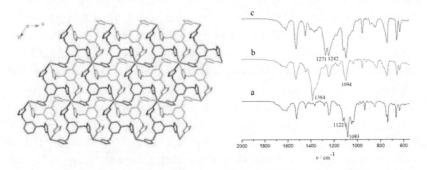

Figure 9. 2D structure of complex **22** (left) and IR spectra of complex **22** (a), and the exchanged products of **22A** (b) and **22B** (c) (right).

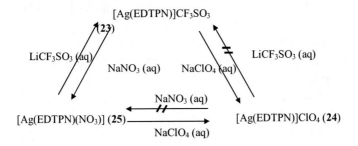

Figure 10. The anion exchanges among the compounds **23**, **24** and **25**.

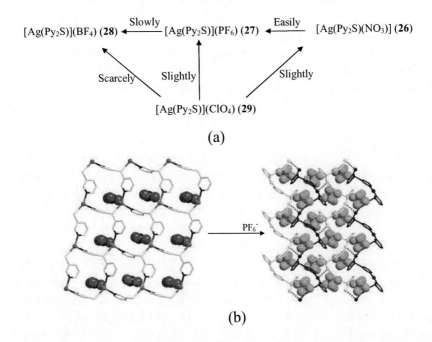

(a)

(b)

Figure 11. (a) Schematic drawing of the anion exchange among complexes **26–28**.
(b) The skeletal structure changes from **26** to **27** upon the anion exchange.

Jung's group investigated coordination complexes obtained from reactions of flexible pyridine-based ligand 3,3'-thiobispyridine (Py$_2$S) with varied silver(I) salts, namely [Ag(Py$_2$S)(NO$_3$)] (**26**) with 2D network structure in which the nitrate anions weakly coordinated to the metal atom, [Ag(Py$_2$S)]PF$_6$ (**27**), [Ag(Py$_2$S)]BF$_4$ (**28**) and [Ag(Py$_2$S)]ClO$_4$ (**29**) with similar infinite helical chain structure and the anions of PF$_6^-$, BF$_4^-$ and ClO$_4^-$ are free of coordination

[52]. The results of the anion exchange are summarized in Figure 11a. The counteranion exchange of **29** with NO_3^- or PF_6^- proceeds slightly, and with ClO_4^- scarcely occurs at room temperature. The PF_6^- in **27** exchanged with BF_4^- is slowly. However, the anion exchange of NO_3^- in 2D network of **26** with PF_6^- takes easily accompanying with the skeletal structure converted from **26** to **27** (Figure 11b), while the reverse exchange of $[Ag(Py_2S)]PF_6$ (**27**) with NO_3^- is not easy [52].

THREE-DIMENSIONAL FRAMEWORKS

Similar to the infinite 1D chains and 2D networks mentioned above, the reported 3D coordination frameworks also show promising properties and potential applications. Particularly, nanoporous metal-organic hybrid frameworks have recently been developed into an important new class of solid state materials due to their potential applications in the fields of storage, separation, catalysis and so on [53 - 56]. The assembly reaction of ligand 2, 4, 6-tris[4-(imidazol-1-ylmethyl)phenyl]-1, 3, 5-triazine (timpt) with nickel(II) perchlorate hexahydrate leads to the formation of a 3D framework $[Ni(timpt)_2](ClO_4)_2$, (**30**) with self-penetration as schematically shown in Figure 12 [57]. The uncoordinated perchlorate anions are located within the voids of the 3D framework, and interestingly the two perchlorate anions have different environments: one is loosely bound to the framework through only one C-H···O hydrogen bond, while the other one is strongly bound to the framework via the formation of four hydrogen bonds. Such difference between the two perchlorate anions leads to the different anion exchange behavior. The results of anion exchange indicate that the ClO_4^- anions in complex **30** can only be partially exchanged by NO_3^- anions. Furthermore, when the anion exchange was carried out by using BF_4^- instead of NO_3^-, and it is the same that only partial anion exchange was observed. This implies that only the perchlorate anion bound to the framework via only one weak hydrogen bond could be exchanged while the one strongly bound to the framework could not be exchanged [57].

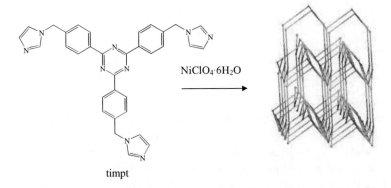

Figure 12. 3D framework of **30** with self-penetration assembled by timpt and NiClO$_4$·6H$_2$O.

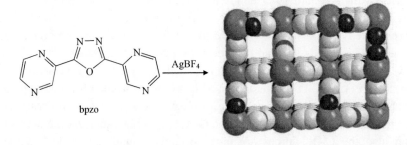

Figure 13. Space-filling view of **31** with square channels obtained by reaction of bpzo and AgBF$_4$.

The partial and selective anion exchange was also observed in a series of PtS type 3D frameworks [Ag(bpzo)]X, [X = BF$_4^-$ (**31**), AsF$_6^-$ (**32**), CF$_3$SO$_3^-$ (**33**), SbF$_6^-$ (**34**)] assembled by the corresponding AgX salts with ligand 2,5-bis(pyrazine)-1,3,4-oxadiazole (bpzo), and a typical example is shown in Figure 13 [58]. It can be seen that the frameworks have 1D square channels with dimension of *ca* 4.1 × 4.1 Å in one direction occupied by counteranions and solvent molecules. The anion exchange property of the frameworks was studied and the results showed that complete and reversible anion exchanges were observed between **31** and **33** as well as **33** and **34**. The largest SbF$_6^-$ anion in **34** could be easily replaced by any one of the other three anions, however, anion exchange of CF$_3$SO$_3^-$ in **33** with BF$_4^-$ to **31** or with CF$_3$SO$_3^-$ to **33** can only partially occur, probably due to the different sizes of the anions. The results suggest that such frameworks may have potential for application as a new kind of molecular-based functional material.

Assembly reaction of cadmium nitrate tetrahydrate with flexible organic ligand 1, 2, 4, 5-tetrakis(imidazol-1-ylmethyl)benzene (tkimb) leads to the formation coordination complex [Cd$_2$(tkimb)$_3$](NO$_3$)$_4$·6H$_2$O (**35**) [59]. The combination of the flexible tetradentate ligands tkimb and six-coordinated Cd(II) atoms results in the formation of a 3D framework structure with 1D channels as illustrated in Figure 14. The NO$_3^-$ anions are filled in the channels through hydrogen bonding interactions, so the framework was expected to exhibit anion exchange property, and the results of IR, elemental analyses and Xray powder diffraction confirmed that the NO$_3^-$ anions in **35** can be completely and reversibly exchanged by ClO$_4^-$.

Kwak and co-workers reported a 3D framework [Cu$_2$(AEP)(OX)](ClO$_4$)$_2$ (**36**) (OX^{2-} = oxalate) obtained via assembly reaction of sodium oxalate, copper nitrate trihydrate, sodium perchlorate and 1-(2-aminoethyl)-piperazine (AEP), which contains 1D nanoscale channels (Figure 15) [60]. The anion exchange between ClO$_4^-$ and PF$_6^-$ from the nanochannels of **36** was investigated in cold water and acetonirile from kinetic and thermodynamic aspects. The results showed that the complete anion exchange from ClO$_4^-$ to PF$_6^-$ took more than 50 days in both solvents, while in the reverse exchange, i.e. from PF$_6^-$ to ClO$_4^-$, about 10 and 12 days were needed for complete anion exchange in water and acetonitrile, respectively [60]. Such observation was interpreted by higher mobility of the ClO$_4^-$ anions in water than in acetonirile, and also the ClO$_4^-$ anions are more strongly bound to the 3D framework through the hydrogen bonding interactions than the of PF$_6^-$ ones. It was concluded that the thermodynamic factor has predominant role in the slow anion exchange [60].

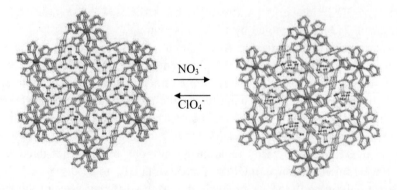

Figure 14. The 3D framework structure of **35** with the nitrate anions filled in the 1D channel (left) and the perchlorate anions (right) upon the reversible anion exchange.

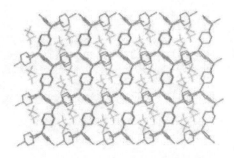

Figure 15. Crystal structure of **36** with perchlorate anions filled in the nanochannels.

In addition to the anion exchange within the cationic 3D frameworks, the cations in the anionic frameworks can also be exchanged by other cations. For example, recently reported 3D frameworks with 1, 3, 5-benzenetricarboxylic acid (H_3btc) and pyridine derivative ligands were found to show cation exchange property [61]. The protonated cationic guests $(HL)^+$ in the 1D anionic channels of the 3D frameworks, $\{(HL)[In_4(OH)_4(btc)_3]\cdot L\cdot 3H_2O\}_n$ [L = pyridine (**37**); 2-picoline (**38**); 4-picoline (**39**)], can be exchanged by K^+, Sr^{2+} or Ba^{2+} ions in full or partial [61]. Furthermore, when the cationic guests at the channels were changed to $(Hdpea)^+$ [dpea = 1, 2-di(4-pyridyl)ethane], it was found the cation exchange is difficult due to the large size of the cationic guests. Therefore, the size of the guest cations plays important role in the exchange. Another anionic 3D framework, $\{[H_2tmdp]_3[In_6(btc)_8]\cdot 40H_2O\}_n$ (**40**) (tmdp = 4,4'-trimethylene dipiperidine), has $In_9(btc)_{11}$ cage-like units as shown in Figure 16 [62]. The protonated cationic guests occupy the multi-intersecting open channels in **40** and the results of cation exchange experiments indicate that the cationic guests can be partially exchanged by K^+, NH_4^+, Ca^{2+}, Sr^{2+} or Ba^{2+} [62].

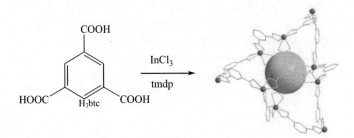

Figure 16. The $In_9(btc)_{11}$ cage-like unit in **40** formed by reaction of H_3btc, tmdp and $InCl_3$.

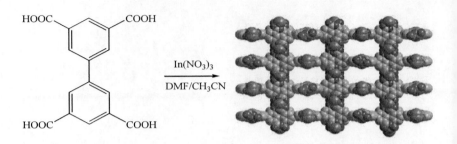

Figure 17. The formation and 3D structure of **41**.

The ion exchange reaction can not only lead to the formation of exchanged product but also affect the property of the exchanged compounds. For example, a 3D framework $(Me_2NH_2)[In(C_{16}H_6O_8)]$ (**41**) was obtained by assembly of biphenyl-3,3′,5,5′-tetracarboxylic acid $(H_4C_{16}H_6O_8)$ and $In(NO_3)_3$ in DMF (N, N-dimethylformamide) and CH_3CN, in which there are square-shaped channels (Figure 17) filled with the $Me_2NH_2^+$ cations [63]. When the compound **41** was immersed in a saturated solution of LiCl for 10 days, the $Me_2NH_2^+$ cations within the pores of the framework were found to be exchanged by half an equivalent of Li^+ and half an equivalent of H_3O^+. It is interesting to find that the hydrogen (H_2) adsorption is enhanced after the cation exchange, namely the Li^+-exchanged 3D framework showed higher hydrogen storage capacity than the $Me_2NH_2^+$ filled framework (**41**) [63].

Further investigation of the H_2 sorption with porous metal-organic materials based on the cation exchange was reported most recently by Eddaoudi et al. [64]. A zeolite-like porous framework $\{[Mg\,(H_2O)_6]^{2+}\}_{24}[In_{48}(HImDC)_{96}]$ (**42**) was obtained by cation exchange from $\{[NMe_2]^+\}_{48}[In_{48}(HImDC)_{96}]$ (**43**) synthesized by reaction of 4,5-imidazoledicarboxylic acid (H_3ImDC) with $In(NO_3)_3 \cdot 2H_2O$ in DMF [64]. A related zeolite-like framework $In_{48}(HImDC)_{96}(HPP)_{24}(DMF)_{36}(H_2O)_{192}$ (**44**) [HPP = 1,3,4,6,7,8-hexahydro-2H-pyrimido(1,2-a)pyrimidine] has been reported previously [65]. The structural analysis showed that each hexaaqua cation $[Mg(H_2O)_6]^{2+}$ is bound to the framework by the formation of hydrogen bonds (Figure 18). The results of the H_2 sorption experiments demonstrate that hydrogen sorption capacity is enhanced upon the cation exchange from $[NMe_2]^+$ to $[Mg(H_2O)_6]^{2+}$ [64].

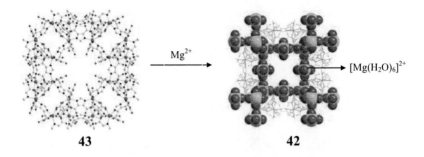

43   **42**

Figure 18. Framework **42** obtained from **43** via Mg^{2+} ion exchange reaction.

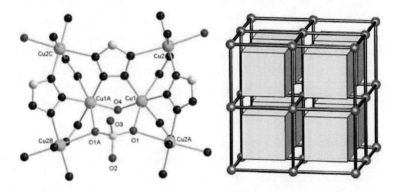

Figure 19. The coordination environment of Cu(II) in compound **45** (left) and 3D framework with **pcu** net (right).

From the reported studies, the ion exchange generally occurs among the guest anions or cations with suitable size and shape so that they can enter and go out the host framework [66], and the interactions between the guest anions/cations and the cationic/anionic framework are also important. In contrast to the guest ion exchange, the coordinated ions is usually difficult for exchange, since such process involves in the rupture of the coordination bond [67] and only limited examples in which the anions weakly coordinated to metal atom are reported to now [46]. However, a recently reported 3D polymer $\{[Cu_4(\mu_2\text{-}OH)(\eta^2\text{:}\eta^2\text{:}\mu_4\text{-}SO_4)(tta)_5]\cdot 3H_2O\}_n$ (**45**) (Htta = tetrazole) with five-nodal (3, 4, 6)-connected **pcu** topology (Figure 19) showed coordinated anion exchange property [23]. When the framework **45** is treated by an aqueous methanolic solution containing Cl^- anions, the coordinated SO_4^{2-} anions could be fully exchanged by Cl^- anions evidenced by elemental analysis, IR spectral measurements, SEM, energy-dispersive Xray spectrometry (EDS) and Xray powder diffraction. Moreover, the results of atomic force microscopy (AFM) revealed that the exchange is solvent mediated rather than solid state transformations [68, 69].

CONCLUSION

In this chapter we outlined four different kinds of metal-organic frameworks with ion exchange property. In general, frameworks with pyridine- or imidazole-containing neutral ligands are cationic and, as a result, the anions serve as counter ions; thus, anion exchange can be expected. In contrast, anionic frameworks constructed by anionic ligands such as multi-carboxylate can show cation exchange property. Therefore, the assembly and structure of the frameworks are important for ion exchange. On the other hand, the dimensionality of the coordination polymers may be increased by employing high coordinate metal ions or multinuclear structural units as the building blocks, since these manifest themselves as highly connected nodes in the polymeric structures. Therefore, rational combination of the multi-bridging ligands and metal salts may, in principle, allow for generation of interesting coordination architectures based on coordination bonding as well as supramolecular interactions.

In conclusion, the design of organic ligands or second building blocks plays crucial rule in the formation of frameworks with specific structure and ion exchange property. Further studies are required for controlling and tuning the framework structure, and then for selective and/or specific ion exchange, which are also tremendously important for crystal engineering and supramolecular chemistry.

REFERENCES

[1]   Batten, SR; Robson, R. *Angew. Chem. Int. Ed*, 1998, 37, 1460-1494.
[2]   Yaghi, OM; Li, H; Davis, C; Richardson, D; Groy, TL. *Acc. Chem. Res.*, 1998, 31, 474-484.
[3]   Blake, AJ; Champness, NR; Hubberstey, P; Li, WS; Withersby, MA; Schröder, M. *Coord. Chem. Rev.*,1999, 183, 117-138.
[4]   Hagrman, PJ; Hagrman, D; Zubieta, J. *Angew. Chem. Int. Ed*, 1999, 38, 2638-2684.

[5]   Moulton, B; Zaworotko, M. J. *Chem. Rev.*, 2001, 101, 1629-1658.

[6]   Seddon, KR; Zaworotko, M. Crystal Engineering: *the Design and Application of Functional Solids*, Kluwer Academic Publishers, Dordrecht, 1999.

[7]   Braga, D; Grepioni, F; Orpen, AG. Crystal Engineering: *from Molecules and Crystals to Materials*, Kluwer Academic Publishers, Dordrecht, 1999.

[8]   Gokel, GW. *Advances in Supramolecular Chemistry*, Vol. 7, JAI Press, Greenwich, 2000.

[9]   Liang, YC; Cao, R; Su, WP; Hong, MC; Zhang, WJ. *Angew. Chem. Int. Ed*, 2000, 39, 3304-3307.

[10]   Zaworotko, MJ. *Chem. Commun*, 2001, 39, 1-9.

[11]   Li, H; Eddaoudi, M; O'Keeffe, M; Yaghi, OM. *Nature*, 1999, 402, 276-279.

[12]   Chui, SSY; Lo, SMF; Charmant, JPH; Orpen, AG; Williams, ID. *Science*, 1999, 283 1148-1150.

[13]   Noro, SI; Kitagawa, S; Kondo, M; Seki, K. *Angew. Chem. Int. Ed*, 2000, 39, 2081-2084.

[14]   Mautner, FA; Cortes, R; Lezama, L; Rojo, T. *Angew. Chem. Int. Ed*, 1996, 35, 78-80.

[15]   Inoue, K; Hayamizu, T; Iwamura, H; Hashizume, D; Ohashi, Y. *J. Am. Chem. Soc.*, 1996, 118, 1803-1804.

[16]   Tong, ML; Chen, XM; Ye, BH; Ji, LN. *Angew. Chem. Int. Ed*, 1999, 38 2237-2240.

[17]   Campos-Fernández, CS; Clérac, R; Dunbar, Kim R. *Angew. Chem. Int. Ed*, 1999, 38, 3477-3479.

[18]   Bu, XH; Tong, ML; Chang, HC; Kitagawa, S; Batten, SR. *Angew. Chem. Int. Ed*, 2003, 43, 192-195.

[19]   Chen, B; Ockwig, NW; Millward, AR; Contreras, DS; Yaghi, OM. *Angew. Chem. Int. Ed*, 2005, 44, 4745-4749.

[20]   Rowsell, JLC; Yaghi, OM. *Micropor. Mesopor. Mater*, 2004, 73, 3-14.

[21]   Rowsell, JLC; Yaghi, OM. *Angew. Chem. Int. Ed*, 2005, 44, 4670-4679.

[22]   Janiak, C. *Dalton Trans.*, 2003, 2781-2804.

[23]   Yang, HY; Li, LK; Wu, J; Hou, HW; Xiao, B; Fan YT. *Chem. Eur. J*, 2009, 15, 4049-4056.

[24]   Kusukawa, T; Fujita, MJ. *J. Am. Chem. Soc.*, 2002, 124, 13576-13582.

[25]   Sun, WY; Kusukawa, T; Fujita, MJ. *J. Am. Chem. Soc.*, 2002, 124, 11570-11571.

[26]   Yoshizawa, M; Takeyama, Y; Okano, T; Fujita, M. *J. Am. Chem. Soc.*, 2003, 125, 3243-3247.

[27]   Yoshizawa, M; Kusukawa, T; Fujita, M; Sakamoto, S; Yamaguchi, K. *J. Am. Chem. Soc.*, 2001, 123, 10454-10459.

[28]   Bourgeois, JP; Fujita, M; Kawano, M; Sakamoto, S; Yamaguchi, K. *J. Am. Chem. Soc.*, 2003, 125, 9260-9261.

[29]   Yoshizawa, M; Tamura, M; Fujita, M. *J. Am. Chem. Soc.*, 2004, 126, 6846-6847.

[30]   Yoshizawa, M; Miyagi, S; Kawano, M; Ishiguro, K; Fujita, M. *J. Am. Chem. Soc.*, 2004, 126, 9172-9173.

[31]   Nakabayashi, K; Kawano, M; Yoshizawa, M; Ohkoshi, SI; Fujita, M. *J. Am. Chem. Soc.*, 2004, 126, 16694-16695.

[32]   Bourgeois, JP; Fujita, M; Kawano, M; Sakamoto, S; Yamaguchi, K. *J. Am. Chem. Soc.*, 2003, 125, 9260-9261.

[33]   Liu, HK; Sun, WY; Ma, DJ; Yu, KB; Tang, WX. *Chem. Commun*, 2000, 591-592.

[34]   Sun, WY; Yoshizawa, M; Kusukawa, T; Fujita, M. *Curr. Opin. Chem. Biol.*, 2002, 6, 757-764.

[35]   Sun, WY; Fan, J; Okamura, T; Xie, J; Yu, KB; Ueyama, N. *Chem. Eur. J*, 2001, 7, 2557-2562.

[36]   Liu, HK; Huang, XH; Lu, TH; Wang, XJ; Sun, WY; Kang, BS. *Dalton Trans.*, 2008, 3178-3188.

[37]   Fochi, F; Jacopozzi, P; Wegelius, E; Rissanen, K; Cozzini, P; Marastoni, E; Fisicaro, E; Manini, P; Fokkens, R; Dalcanale, E. *J. Am. Chem. Soc.*, 2001, 123, 7539-7552.

[38]   Noro, S; Kitaura, R; Kondo, M; Kitagawa, S; Ishii, T; Matsuzaka, H; Yamashita, M. *J. Am. Chem. Soc.*, 2002, 124, 2568-2583.

[39]   Pan, L; Woodlock, EB; Wang, XT; Lam, KC; Rheingold, AL. *Chem. Commun*, 2001, 1762-1763.

[40]   Yaghi, OM; Li, HL; Groy, TL. *Inorg. Chem.*, 1997, 36, 4292-4293.

[41]   Qiu, YC; Liu, ZH; Li, YH; Deng, H; Zeng, RH; Zeller, M. *Inorg. Chem.*, 2008, 47, 5122-5128.

[42]   Wang, P; Ma, JP; Dong, YB; Huang, RQ. *J. Am. Chem. Soc.*, 2007, 129, 10620-10621.

[43]   Yaghi, OM; Li, H. *J. Am. Chem. Soc.*, 1995, 117, 10401-10402.

[44]   Yaghi, OM; Li, H. *J. Am. Chem. Soc.*, 1996, 118, 295-296.

[45]   Hoskins, BF; Robson, R. *J. Am. Chem. Soc.*, 1990, 112, 1546-1554.

[46]   Min, KS; Suh, MP. *J. Am. Chem. Soc.*, 2000, 122, 6834-6840.

[47]   Fan, J; Gan, L; Kawaguchi, H; Sun, WY; Yu, KB; Tang, WX. *Chem. Eur. J*, 2003, 9, 3965-3973.

[48]   Fan, J; Sun, WY; Okamura, T; Tang, WX; Ueyama, N. *Inorg. Chem.*, 2003, 42, 3168-3175.

[49]   Zhao, W; Fan, J; Song, Y; Kawaguchi, H; Okamura, T; Sun, WY;

Ueyama, N. *Dalton Trans*, 2005, 1509-1517.

[50] Fan, J; Sui, B; Okamura, T; Sun, WY; Tang, WX; Ueyama, N. *J. Chem. Soc., Dalton Trans.*, 2002, 3868-3873.

[51] Xu, GC; Ding, YJ; Huang, YQ; Liu, GX; Sun, WY. *Micropor. Mesopor. Mat*, 2008, 113, 511-522.

[52] Jung, O; Kim, YJ; Lee, Y; Chae, HK; Jang, HG; Hong, JK. *Inorg. Chem.*, 2001, 40, 2105-2110.

[53] Dybtsev, DN; Chun, H; Kim, K. *Angew. Chem. Int. Ed*, 2004, 43, 5033-5036.

[54] Hollingsworth, MD. *Science*, 2002, 295, 2410-2413.

[55] Eddaoudi, M; Kim, J; Rosi, N; Vodak, D; Wachter, J; O'K eeffe, M; Yaghi, OM. *Science*, 2002, 295, 469-472.

[56] Wright, AT; Anslyn, EV. *Chem. Soc. Rev.*, 2006, 35, 14-28.

[57] Wan, SY; Huang, YT; Li, YZ; Sun, WY. *Micropor. Mesopor Mat*, 2004, 73, 101-108.

[58] Du, M; Zhao, XJ; Guo, JH; Batten, SR. *Chem. Commun*, 2005, 4836-4838.

[59] Xu, GC; Hua, Q; Okamura, T; Bai, ZS; Ding, YJ; Huang, YQ; Liu, GX; Sun, WY; Ueyama N. *CrystEngComm.*, 2009, 11, 261-270.

[60] Chang, M; Chung, M; Lee, BS; Kwak, CH. *J. Nanosci. Nanotechnol*, 2006, 6, 3338-3342.

[61] Lin, ZZ; Chen, L; Yue, CY; Yuan, DQ; Jiang, FL; Hong MC. *J. Solid State Chem.*, 2006, 179, 1154-1160.

[62] Lin, ZZ; Jiang, FL; Chen, L; Yue, CY; Yuan, DQ; Lan, AJ; Hong, MC. *Cryst. Growth Des*, 2007, 7, 1712-1715.

[63] Yang, SH; Xiang, L; Blake, AJ; Thomas, KM; Hubberstey, P; Champness, NR; Schröder, M. *Chem. Commun*, 2008, 6108-6110.

[64] Nouar, F; Eckert, J; Eubank, JF; Forster, P; Eddaoudi, M. *J. Am. Chem. Soc.*, 2009, 131, 2864-2870.

[65] Liu, Y; Kravtsov, VC; Larsen, R; Eddaoudi, M. *Chem. Commun*, 2006, 1488-1490.

[66] Choi, HJ; Suh, MP. *Inorg. Chem.*, 2003, 42, 1151-1157.

[67] Wang, Y; Cheng, P; Song, Y; Liao, DZ; Yan, SP. *Chem. Eur. J*, 2007, 13, 8131-8138.

[68] Pina, CM; Fernandez-Diaz, L; Prieto, M; Putnis, A; *Geochim. Cosmochim. Acta*, 2000, 64, 215-221.

[69] Malkin, AJ; Kuznetsov, YG; McPherson, A. *J. Cryst. Growth*, 1999, 196, 471-488.

In: Chemical Crystallography   ISBN: 978-1-60876-281-1
Editors: Bryan L. Connelly, pp. 153-159 © 2010 Nova Science Publishers, Inc.

Chapter 5

SUBSTITUENT EFFECT ON THE STRUCTURES OF ZINC 1,4-BENZENEDICARBOXYLATE COORDINATION POLYMERS SYNTHESIZED IN DIMETHYL SULFOXIDE

Shi-Yao Yang [*] *and Xiao-Bin Xu*

College of Chemistry and Chemical engineering,
Xiamen University, Xiamen, 361005, China.

ABSTRACT

The coordination polymer $[Zn(tmbdc)(dmso)_2] \cdot 2(DMSO)$ (tmbdc = 2,3,5,6-tetramethyl-1,4-benzenedicarboxylate) has been synthesized by layer diffusion in DMSO (dimethyl sulfoxide) solution. The compound contains 1D chain formed by octahedraly coordinated Zn^{2+} ion chelated by the carboxyl groups of tmbdc. In another recently reported coordination polymer $[Zn_2(bdc)_2(dmso)_2] \cdot 5(DMSO)$ (bdc = 1,4-benzenedicarboxylate) prepared under the same condition, pairs of Zn^{2+} ions are bridged by four carboxyl groups to form paddle-wheel sub unit and the 2D (4,4) net structure. Analysis of the structures reveals that the substituents of the ligands determine the coordination environments of zinc ions and the coordination modes of the carboxyls, and thus the final structures of the coordination polymers.

[*] Corresponding author: Email: syyang@xmu.edu.cn.

INTRODUCTION

The properties and potential applications of coordination polymers (CPs) are closely related to their structures, therefore the rational design and synthesis of CPs is one of the most important and urgent task of crystal engineering. [1-4] However, it is still a great challenge to synthesize a structure by design because there are numerous influences such as the substituents on ligands, solvents, synthetic temperatures, etc, that can have decisive roles in determining of the structure and crystal packing. [4-13].

These uncertainties can be reduced by the use of well designed ligands that bind metal ions at chelating sites, such as carboxylates linkers that have the ability to aggregate metal ions into secondary building units (SBUs). Among the organic ligands used in the assembly of coordination polymers, benzene polycarboxylate species have been extensively explored to produce robust crystalline materials due to their versatile coordination ability. [1,4] Some analogous compounds possessing substituents, such as aryl, nitro or hydroxyl groups, may provide new chance to rationally modulate and control the network structures and new understanding of crystal engineering. In particular, when there are different substituents on ligands, the coordination environments of metal ions can be changed, and the structures with completely different topologies are obtained. [5-9]

The influence of the four methyl groups of H_2tmbdc (2,3,5,6-tetramethyl-1,4-benzenedicarboxylic acid) on the assembly of CPs in DMF (N,N-dimethylformamide) has been investigated. In the work of Yaghi's group, they found that the square grid structure constructed from paddle-wheel units of Zn^{2+} could not be formed with tmbdc in MOF-47 because of the sterically demanding. They claimed that the dihedral angle (Θ, Scheme 1) between the planes of benzene and carboxyl groups play a determining role in the formation of the paddle-wheel motif. The Θ value for MOF-47 was 84°, and a tetrahedral SBU and therefore a double layer motif of the structure was formed. [5] Nevertheless, in the work of Kim's group, the substituents did not show influence on the formation of the paddle-wheel structure in [Zn_2(1,4-bdc)(tmbdc)(dabco)] (Θ = 75.0(1)°), and [Zn_2(tmbdc)$_2$(dabco)] (Θ = 73.9(1)°) (dabco = 4-diazabicyclo[2.2.2]octane). [4] The contradiction shows that other factors may also play the decisive roles. In many cases, it is difficult to predict which factor will be the dominant one. The true engineering of crystal structures is still a distant aim which demands more extensive and intensive investigations.

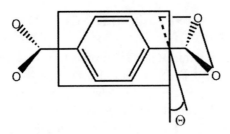

Scheme 1. Definition of the dihedral angle, Θ, between carboxyl and benzene ring.

Among the other factors, the effect of solvent on the assembly of CPs is apparent. The solvents, in many cases act as ancillary ligands, can compete with the ligands. The coordination environment adopted by the metal centre, and therefore the structure of the coordination polymer, can be very variable and unpredictable. [10-13] DMF and DMSO (dimethyl sulfoxide) are among the solvents that have great tendencies to be included in organic crystals *via* multi-point recognition between solvent and solute. [14] DMF has been largely used in the syntheses of CPs, especially for highly porous structures, while DMSO is much less explored. According to CCDC search, [15] there are 56 zinc benzenecarboxylate CPs which have been synthesized with DMF as solvent, and for DMSO, the number is 11. In particular, for zinc 1,4-benzenedicarboxylates, the number for DMF is 21, and for DMSO, only 3: $[Zn_4(OH)(bdc)_3(dmso)_4]\cdot2H_2O$ [16,17], $[Zn_2(bdc)_2(dmso)_2]\cdot5DMSO$ **1** [18], and $(Bu_4N)[Zn(bdc)_{1.5}(dmso)]\cdot0.67DMSO\cdot0.25pz$ (pz = pyrazine) [19].

We have synthesized and characterized the zinc 1,4-benzenedicarboxylate coordination polymer $[Zn_2(bdc)_2(dmso)_2]\cdot5DMSO$ **1** by layer diffusion in DMSO. [18] In **1**, pairs of Zn^{2+} ions are bridged by four carboxyl groups to form paddle-wheel SBU. (Fig. 1 a) The SBUs are further extended by bdc and result in the formation of the 2D (4,4) net. The 2D nets pack along the *a* axis forming 1D channels occupied by large amount of solvent DMSO molecules. (Fig. 2 a) The ratio of solvent molecules to Zn^{2+} ions is 2.5, which is much higher than 0~1 for similar structures synthesized in DMF or analog solvents. [3,10,13]

In order to explore the substituent effect on the assembly of coordination polymers, the influence of other factors should be excluded. Therefore the same synthetic condition as that for **1** was applied to synthesize the zinc 2,3,5,6-tetramethyl-1,4-benzenedicarboxylate, $[Zn(tmbdc)(dmso)_2]\cdot2(DMSO)$ **2**, except that H₂bdc was replaced by H₂tmbdc. [20]

RESULT AND DISCUSSION

X-ray structure determination [21] reveals that there are one Zn^{2+} ion, one tmbdc dianion and four DMSO molecules in the asymmetric unit in **2**. The Zn^{2+} ion is in a severely distorted octahedral geometry, coordinated by two carboxyl groups in chelate mode and two DMSO molecules. (Fig. 1 b) Due to the steric hindrance of the four methyl groups, the Θ angle in **2** is 85.4°. This value is even larger than 84° in MOF-47, [5] and the paddle-wheel SBU is not adopted by **2**, either. The orientations of the two tmbdc dianions coordinated to the same Zn^{2+} ion are at an angle of 134.5°, therefore a 1D chain is formed. (Fig. 2 b) By comparison with **2**, the average Θ in **1** is 10.55 (23.0, 7.5, 3.5 and 8.2° for four crystallographically independent carboxyl groups, all are smaller than 25° in $Zn(ABDC)(DMF)\cdot(C_6H_5Cl)_{0.25}$ (MOF-46, ABDC = 2-amino-1,4-benzenedicarboxylate) [5]). The methyl groups play a determining role in the formation of the mononuclear metal center and thus the 1D structure of **2**. There are also DMSO solvent molecules included in the crystal structure of **2**. It is worth noticing that the ratio of solvent DMSO molecules to Zn^{2+} ions is 2, which is unexpected high for closely packed 1D chain structure. The result again shows the strong tendency of DMSO to be included in the structures of coordination polymers. Compared to DMF, DMSO also has a stronger tendency to coordinate to Zn^{2+} ion.

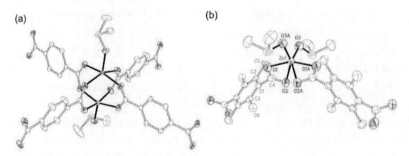

Figure.1. The coordination environments of Zn^{2+} ions in **1** (a) and **2** (b). Hydrogen atoms and solvent DMSO molecules are omitted for clarity. Selected bond lengths (Å) and angles (°) for **2**: Zn1–O1 2.043(3), Zn1–O2 2.346(3), Zn1–O3 2.022(3), O1–Zn1–O2 58.68(12), O1–Zn1–O3 102.62(14), O1–Zn1–O1A 156.9(2), O1–Zn1–O2A 104.12(13), O1–Zn1–O3A 91.85(13), O2–Zn1–O3 89.84(14), O2–Zn1–O2A 92.79(18), O2–Zn1–O3A 150.14(12), O3–Zn1–O3A 102.4(2). Symmetry code: A −x+1, y, −z+3/2.

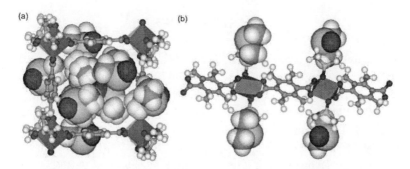

Figure 2. Views of the structures of **1** (a) and **2** (b). The coordination polymers are shown in ball and stick mode, zinc ions are shown in polyhedra, DMSO solvent molecules are shown in CPK mode.

CONCLUSIONS

In conclusion, 1D coordination polymer [Zn(tmbdc)(dmso)$_2$]·2(DMSO) **2** has been synthesized with 2,3,5,6-tetramethyl-1,4-benzenedicarboxylic acid in DMSO. The structure contains 1D chains formed by octahedraly coordinated Zn^{2+} ions chelated by the carboxyl groups of tmbdc, rather than the 2D (4,4) nets constructed from paddle-wheel SBU of pairs Zn^{2+} ions as found in [Zn$_2$(bdc)$_2$(dmso)$_2$]·5DMSO **1**. Analysis of the structure reveals that the steric hindrance of the four methyl groups of tmbdc determines the coordination environments of the zinc ions and the coordination modes of the carboxyls, and thus the final structures of the coordination polymers. The result also shows that DMSO is a stronger ancillary ligand and is also easier to be included in the structures of coordination polymers, compared to DMF. DMSO can be a better solvent for the syntheses of porous coordination polymers.

ACKNOWLEDGEMENT

We thank the National Natural Science Foundation of China (20471049) for financial assistance.

REFERENCES

[1]   Eddaoudi, M; Kim, J; Rosi, N; Vodak, D; Wachter, J; O'Keeffe, M; Yaghi, O. M. *Science*, 2002, 295, 469-472.

[2]   Rowsell, JLC; Millward, AR; Park, KS; Yaghi, OM. *J. Am. Chem. Soc.*, 2004, 126, 5666-5667.

[3]   Li, H; Eddaoudi, M; Groy, TL; Yaghi, OM. *J. Am. Chem. Soc.*, 1998, 120, 8571-8572.

[4]   Chun, H; Dybtsev, DN; Kim, H; Kim, K. *Chem. Eur. J*, 2005, 11, 3521-3529.

[5]   Braun, ME; Steffek, CD; Kim, J; Rasmussen, PG; Yaghi, OM. *Chem. Commun*, 2001, 2532-2533.

[6]   Li, XJ; Cao, R; Guo, ZG; Li, YF; Zhu, XD. *Polyhedron*, 2007, 26, 3911-3919.

[7]   Du, M; Zhang, ZH; You, YP; Zhao, XJ. *CrystEngComm*, 2008, 10, 306-321.

[8]   Ren, H; Song, TY; Xu, JN; Jing, SB; Yu, Y; Zhang, P; Zhang, LR. *Cryst. Growth Des.*, 2009, 9, 105-112.

[9]   Ma, LF; Wang, LY ; Wang, YY; Du, M; Wang, JG. *CrystEngComm*, 2009, 11, 109-117.

[10] Edgar, M; Mitchell, R; Slawin, AMZ; Lightfoot, P; Wright, PA. *Chem. Eur. J*, 2001, 7, 5168-5175.

[11] Hawxwell, SM; Adams, H; Brammer, L. *Acta Cryst*, 2006, B62, 808-814.

[12] Burrows, AD; Cassar, K; Friend, RMW; Mahon, MF; Rigby, SP; Warren, JE. *CrystEngComm*, 2005, 7, 548-550.

[13] Wang, FK; Yang, SY; Huang, RB; Zheng, LS; Batten, SR. *CrystEngComm*, 2008, 10, 1211-1215.

[14] Nangia, A; Desiraju, GR. *Chem. Commun*, 1999, 605-606.

[15] Cambridge Structural Database; Version 5.30, update Feb 2009, see also Allen, FH. *Acta Cryst*, 2002, B58, 380-388.

[16] Wang, R; Hong, M; Liang, Y; Cao, R. *Acta Cryst*, 2001, E57, m277-m279.

[17] Zevaco, TA; Männle, D; Walter, O; Dinjus, E. *Appl. Organometal. Chem.*, 2007, 21, 970-977.

[18] Yang, SY; Long, LS; Huang, RB; Zheng, LS; Ng, SW. *Acta Cryst*, 2005, E61, m1671-m1673.

[19] Yang, SY; Du, C; Huang, RB; Ng, SW. *Acta Cryst*, 2007, E63, m2788.

[20] Synthesis of 2: Zinc nitrate hexahydrate (0.059 g, 0.20 mmol) in DMSO (2 ml) was placed in a 0.5 mm × 15 mm test tube, and then 2,3,5,6-tertbutyl-1,4-benzenedicarboxylic acid (0.052 g, 0.20 mmol) in DMSO (2 ml) was carefully put on top of it, finally, tributylamine (0.20 ml) was added on the top. The tube was covered and set aside for five days. Colorless needle-shaped crystals (0.066g, yield 55%) formed on the wall of the tube. Elemental analysis, $C_{20}H_{36}O_8S_4Zn$: found (calc.) C 40.61 (40.13), H 5.88 (6.02), S 21.17 (21.44)%.

[21] Crystal data for [Zn(tmbdc)(dmso)$_2$]·2DMSO: Formula $C_{20}H_{36}O_8S_4Zn$, M = 598.10, monoclinic, space group $C2/c$, a = 12.1980(9), b = 12.9463(10), c = 17.8471(13) Å, β = 102.705(1)°, V = 2749.4(4) Å^3, Z = 4, D_c = 1.445 Mg/m^3, μ = 1.237 mm^{-1}, T = 223(2) K, R_{int} = 0.0302, full-matrix least-squares refinement on F^2, R_1 = 0.0687, wR_2 = 0.1820, ρ = 0.874, –0.610 e. Å^{-3}.